SAFETY ENGINEERING

NEW DIMENSIONS IN ENGINEERING

Editor
RODNEY D. STEWART

SYSTEM ENGINEERING MANAGEMENT
Benjamin S. Blanchard

NONDESTRUCTIVE TESTING TECHNIQUES
Don E. Bray (Editor)
Don McBride (Editor)

LOGISTICS ENGINEERING
Linda L. Green

NEW PRODUCT DEVELOPMENT: DESIGN AND ANALYSIS
Ronald E. Kmetovicz

**INDEPENDENT VERIFICATION AND VALIDATION:
A LIFE CYCLE ENGINEERING PROCESS FOR QUALITY SOFTWARE**
Robert O. Lewis

DESIGN TO COST
Jack V. Michaels
William P. Wood

**OPTIMAL INVENTORY MODELING OF SYSTEMS:
MULTI-ECHELON TECHNIQUES**
Craig C. Sherbrooke

COST ESTIMATING, SECOND EDITION
Rodney D. Stewart

PROPOSAL PREPARATION, SECOND EDITION
Rodney D. Stewart
Ann L. Stewart

SAFETY ENGINEERING
James CoVan

SAFETY ENGINEERING

James CoVan

A Wiley-Interscience Publication
JOHN WILEY & SONS, INC.
New York • Chichester • Brisbane • Toronto • Singapore

This text is printed on acid-free paper.

Library of Congress Cataloging in Publication Data:

CoVan, James, 1940–
    Safety engineering / James CoVan.
      p. cm. — (New dimensions in engineering)
    Includes index.
    ISBN 0-471-55612-2
    1. Industrial safety. I. Title. II. Series.
T550.C68 1994
620.8′6—dc20   93-2777

Printed in the United States of America

10 9 8 7 6 5 4 3 2 1

To the memory of my father, Jack P. CoVan, formerly an industrial engineering professor emeritus,

and

To Ralph J. Vernon, Ph.D., C.I.H. and C.S.P., retired industrial engineering professor and active consultant.

Both provided inspiration and guidance that have not been forgotten. These are but two of numerous people who have contributed to the experiences necessary to write to such a broad, complicated, yet important subject. It was only a start to capture the essence of an elusive profession.

CONTENTS

Foreword   xi

List of Acronyms   xiii

1 Introduction   1

Background and Development   1

Basic Concepts   2

2 Safety Terminology   9

Safety   9

Other Related Terms   10

3 The Engineering Approach   13

The Safety Dichotomy   13

4 Safety Skills   19

Accident Investigation   20

Agricultural Safety   27

Back Injury   30

Catastrophes   37

Confined Space Safety   40

Construction Safety   42

Contractor Safety   47

Cryogenic Safety   49

Electrical Safety   51

Emergency Response   54

Ergonomics   62

Facility Safety   68

Fall Protection   71

Fire Safety   74

Hazardous Materials   79

Hazardous Operations   81

Industrial Hygiene   86

Laser Safety   96

Life Safety   100

Lockout/Tagout   103

Materials Handling Safety   106

Mining Safety   107

Off-the-Job Safety   108

Office Safety   114

OSHA   117

Personal Protective Equipment   122

Pressure Systems   131

Public Safety   133

Robotics Safety   138

Safety Training   141

Safety Trends   145

Software Safety   149

Structural Integrity Loss   153

System Safety   154

Transportation Safety   166

Walking and Working Surfaces   170

Waste Handling   171

Welding Safety   174

5 Safety Management Functions and Programs   **177**

Management Functions   178

Functions of the Safety Professional   180

Safety Programs   182

Conclusion   186

6 Safety Documentation   **187**

Specific Safety Documentation   187
Analysis of Documentation   188
Change Management   190

7 Safety References   **193**

Computerization   193
Electronic Databases   198
Safety Texts   198
Safety Journals   198
Trade Magazines and Digests   199
Safety Databanks   199
Safety Apocrypha   199

8 Case Studies of Safety Engineering Applications   **201**

Case I   201
Case II   203
Case III   205
Suggested Safety Engineering Application: Answers to
   the Case Studies   206

References   213

Index   227

FOREWORD

Safety engineering is an increasingly important and growing "horizontal" dimension of engineering that cuts across all of the traditional "vertical" dimensions (civil, mechanical, electrical, chemical, and software engineering). Jim CoVan has prepared this volume in the New Dimensions in engineering series for the professional who wants an overview of the fundamentals and an insight into the subtleties of this expanding discipline. Whether you are an industrial safety engineer, a systems safety engineer, or a manager of any of the many facets of high technology products, you will be informed and enlightened by his presentation for making commercial, industrial, and government projects safer.

As projects become more complex and more potent, the consequences of their malfunction become more serious. Lack of care in manufacturing, testing, or operation can cause substantial losses in productive working time, in personnel health and protection, and in profits. Legal issues alone can paralyze a company or organization and can result in enormous expenditures. A well-planned, organized, systematic, meticulously carried out safety program can be the most significant cost avoidance measure available to any organization. But it must be implemented before, rather than after, the big catastrophe. The principles in this book, if conscientiously applied, can prevent the devastating effects of improper or unsafe practices in the creation and delivery of work outputs or work activities.

We commend this book to you as a vitally important volume in your technical and management library. We are convinced that you will not only become more well informed of the safety principles needed for efficient operations, but will enjoy Jim CoVan's spicy, direct style and insightful perspective.

RODNEY D. STEWART
Series Editor

LIST OF ACRONYMS

ACGIH	American Conference of Governmental Industrial Hygienists
ADA	Americans With Disabilities Act
AFSCs	Air Force specialty codes
AICHE	American Institute of Chemical Engineers
AIHAJ	American Industrial Hygiene Association Journal
ANSI	American National Standards Institute
ASSE	American Society of Safety Engineers
ASTM	American Society for Testing and Materials
BBC	Basic Building Code
BLEVE	boiling liquid expanding vapor explosion
BLR	Business & Legal Reports, Inc.
BLS	Bureau of Labor Statistics
BNA	Bureau of National Affairs, Inc.
CAP	Cooperative Assessment Program
CCH	Commercial Clearing House, Inc.
CDRH	Federal Drug Agency Center for Devices and Radiological Health
CET	corrected effective temperature
CM	change management
CPR	cardiopulmonary resuscitation
CPSA	Consumer Product Safety Act
CPSC	Consumer Product Safety Commission
CSP	certified safety professional
CTD	cumulative trauma disorder
CTS	Carpal Tunnel Syndrome
CW	continuous wave
dB	decibel
DR	data requirements
DWI	driving while intoxicated

ECC	European Economic Community
EMF	electromagnetic field
EMI	electromagnetic interference
EMS	emergency medical services
EPA	Environmental Protection Agency
EPRI	Electrical Power Research Institute
ER	emergency response
EVA	extra-vehicular activity
FAS	fall arresting system
FHWA	Federal Highway Administration
FR	flame retardant
FTA	fault tree analysis
GFCIs	ground fault current interrupters
GULHEMP	General physique, Upper extremities, Lower extremities, Hearing, Eyesight, Mentality, and Personality
HAZOP	hazard and operability study
HazOps	hazardous operations
HIPO	high potential and critical incident techniques
HLDI	Highway Loss Data Institute
HRA	human reliability analysis
HSI	Heat Stress Index
HVAC	heating, ventilation and air conditioning
IH	industrial hygiene
IIHS	Insurance Institute for Highway Safety
ISEA	Industrial Safety Equipment Association
JSA	job safety analysis
KISS	keep it simple stupid
Laser	light amplification by stimulated emission of radiation
LCG	liquid-cooled garment
MCASP	Motor Carrier Safety Assistance Program
MES	multilinear events sequence
MESA	Mine Enforcement and Safety Administration
MMH	manual material handling
MORT	management risk and oversight tree
MSDS	material safety data sheet
MSHA	Mine Safety Health Administration
MVR	motor vehicle record
NASA	National Aeronautics and Space Administration
NBC	National Building Code
Nd:YAG	neodymium yttrium aluminum garnet
NEISS	National Electronics Injury Surveillance System
NFPA	National Fire Protection Association

NHTSA	National Highway Traffic Safety Administration
NHZ	nominal hazard zone
NIOSH	National Institute for Occupational Safety and Health
NSC	National Safety Council
NTSB	National Transportation Safety Board
O&SHA	operating and support hazard analysis
OHA	operating hazard analysis
OJT	on-the-job training
OSH Act	Occupational Safety and Health Act
OSHA	Occupational Safety and Health Administration
OTJ	off-the-job safety
PELs	permissible exposure levels
PHA	preliminary hazard analysis
PRA	probabilistic risk assessment
PRACA	problem reporting and corrective action
PSF	performance shaping factors
R-E-C	Recognition-Evaluation-Control
RCRA	Resource Conservation and Recovery Act
RFI	radio frequency interference
RICE	rest, ice, compression, and elevation
RMS	remote manipulating systems
ROPS	rollover protection system
RPE	registered professional engineer
RPTARS	reliability/problem trend analysis and reporting system
SARA	Superfund Amendments and Reauthorization Act
SBC	Southern Building Code
SHA	system hazard analysis
SSHA	subsystem hazard analysis
THERP	technique for human error rate prediction
TLVs	threshold limit values
TSCA	Toxic Substance Control Act
UBC	Uniform Building Code
VCE	vapor cloud explosion
VDT	video display terminal
VDU	video display unit
VLF	very low frequency
VWF	vibration-induced white finger (formerly Raynaud's phe-nomena
W/WS	walking and working surfaces
WBGT	wet bulb glove temperature

SAFETY ENGINEERING

1
INTRODUCTION

This book is not intended to be a safety handbook or a cookbook for implementing safety. Safety is a broad, multidisciplinary topic and this book addresses the engineering aspects of the subject. As an introductory text, reasons for safety applicability to a variety of subjects are given, general methods of application of the safety engineering discipline are presented, and suggestions for further information are offered. As with any text on a broad subject, compromises have been made to keep the information to a reasonable size, while attempting to emphasize the important points. And as with any treatment of a discipline, biases are introduced according to the background of the author. Admitted bias is given to engineering versus behavioral aspects. Systems engineering is the preferred underpinning of safety engineering. Management is included as a valid part of engineering practice. Finally, there is strong preference bias given to practical aspects over the theoretical. For instance, in this book, safety is defined as a relative acceptability of losses, rather than the more common (and improper in my opinion) absence of hazard or risk.

Some safety engineers and safety engineering authors will disagree with many of the conclusions and premises here. Students of this text are encouraged to think for themselves, weighing information in context of the whole. Correctness of safety engineering may not be determined by survey of writings; it must be determined by scientific analysis and critical thought.

BACKGROUND AND DEVELOPMENT

Safety engineering is a diverse and often poorly understood subject. Many critics doubt that engineering is involved. Yet industry and government have recognized the value of controlling monetary and physical losses, so much so that legislation has required various safety-related actions for many years.

A weakness of the discipline that developed to serve the underlying need is that many of its practitioners were untrained and undisciplined in its application. Many organizations cast safety functions into a mold of the injury-crippled, over-the-hill, minimal performer, or catch-all personnel function. Few persons from the discipline ever rose to executive status without abandoning the discipline for operational or engineering jobs. Few jobs were qualified with professional requirements. It took a long time for companies to make the transition from the "safety man" concept to the "safety manager" concept. This text emphasizes both the engineering- and management-based professional aspects of safety.

As the monetary effects of losses rose in proportion to increasing legislation and legal pressure, the safety discipline as a whole began to bootstrap itself and develop wider professionalism. Safety and health legislation is important to the development of the profession of safety engineering but a detailed treatment of this subject is too detailed for inclusion here. Many references are available that discuss safety laws from biblical to present time. Deuteronomy 19:5 discusses an axe head flying off when cutting wood, accidentally killing someone, and the perpetrator being allowed to flee to a city of refuge to avoid death. Translation of the code of the Babylonian king Hammurabi (about 2200 B.C.) revealed penalties for house-collapse death resulting from poor construction. Between biblical times and the present, the rate of safety legislation has continued to grow and is nearing an exponential growth rate.

Safety engineering is a relatively recent phenomenon. Engineering disciplines were pioneered in the seventeenth and eighteenth centuries. Early uses of engineering pertained to construction of engines of war and associated works. As it is now known, "engineering" involves creative application of scientific principles to design, develop, construct, and operate systems for an intended use *with safety to life and property*. From a definitional stance, safety engineering is inherent in all engineering. Safety engineering may be thought of as application of engineering principles to the subject of safety. It is important to recognize that "safety" in a broad context, refers to both safety and health. This occurs, in part, because of overlap. For our purposes, safety engineering shall be applied to both safety and health. Safety shall include health unless specifically noted otherwise. Safety will be carefully defined later.

BASIC CONCEPTS

The interdisciplinary nature of safety engineering is one of its most fundamental aspects. Figure 1-1 shows that its application requires integration with many other functions and their implementing organizations. Furthermore, to

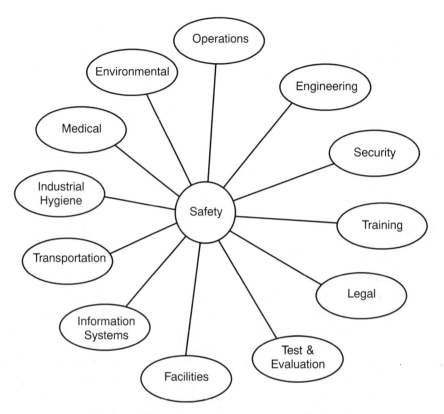

FIGURE 1-1 Safety Interactions

be effective, it must be emphasized that the best application of safety is early in system development, rather than after the fact.

Effective safety programs are often based upon logical, engineered approaches. There are standards of performance that are formally required of all organizational units. These include plans, budgets, staffs, and organizational units. These signify that safety is a manageable activity. This also means that safety cannot be defined as something which is completely random, without recognizable causes, or is a dependent upon an act of God.

Safety is a discipline which requires identification and control of causal agents, in order to provide value to organizations. The safety value is conservation of limited personnel and other resources. If this is done with great leverage (in terms of use of resources), and if it is done optimally within system constraints, safety is successful in contributing to the bottom line. Safety should be carefully integrated into all the functions of an organization, and should be led by safety professionals.

Safety responsibilities of safety professionals and others with safety functions should be identified, measured, and held to account throughout an organization, starting at the top. Organizational safety goals and objectives should be established using normal management practices. Safety goals and objectives are discussed more fully later.

Professionalism

More and more, organizations are requiring professionalism of their safety staffs. Since large sums of money may be saved by good safety engineering, more powerful safety technology has been developed to achieve better results. Just as the traditional professions of engineering and science have strengthened professionalism through training and certification, safety engineering is now achieving a higher level of professionalism with nationally recognized qualifications and certifications. In particular, the Board of Certified Safety Professionals (BCSP) and the American Board of Industrial Hygiene (ABIH) certifications have gained prominence. Also, professional organizations such as the American Society of Safety Engineers (ASSE) and the American Industrial Hygiene Association (AIHA), and many others are growing rapidly.

Many professionals belong to both. It is important to note that certification is not a direct function of the ASSE or AIHA, but independent certifying organizations (BCSP and ABIH) administer the certification process. Nor is certification required for membership in these professional organizations, but it represents the highest recognizable mark of professionalism and is promoted by both professional organizations. Certification is described in more detail later.

Safety Costs

Safety costs should be justified like other forces competing for limited organizational resources. The time value of money should be considered in the planning of safety expenditures. An early expenditure of safety engineering dollars can often avoid much higher loss costs in the future. Safety should be considered to be a long-term investment.

Total 1985 cost of work accidents in the United States was $37.3 billion, triple from that in 1972, according to the National Safety Council (NSC). Yet the 1985-industrial investment in safety and health, $6.8 billion, only doubled over the same time period. (Talty and Walters, 1987). Spread out over the 75 million individuals in the U.S. work force, the $44.1 million total equates to about $590 per worker, per year. Ignoring the finer points of what the costs

were based upon, the point to be made is that the loss costs are ver
icant if they are considered to be funded in great part out of profits. With an
approximate 10% profit margin, businesses would have to generate about
$6,000 per year just to cover average loss costs.

In my opinion, reasonable costs of safety for a program are roughly
10–15% of total program costs for programs with significant risks. One
powerful argument for safety program cost expenditures is the preceding
point that loss costs generally come directly from profit. This means that it
is quite cost effective to avoid costs that must be paid for with large amounts
of sales.

Typical factors contributing to safety costs, other than the costs of the
program itself, include:

- Insurance costs, especially higher rates
- Legal costs
- Regulatory fines
- Abatement costs
- Advertising costs to regain market share lost from adverse safety-related
  publicity
- Redesign/reengineering costs

Safety Philosophy

Safety has always been something of a balancing act. The balance is between
accomplishment of something versus losing the capability to do something in
the process. In this respect, safety is a necessary trade-off that excludes some
of the available methods of doing things. The biggest misconception about
this field is that the term safety is synonymous with perfection. Keep in mind
that some utopian-minded people think of safety in absolute terms, such as
the absence of accidents/losses. Rather, safety is a compromise that optimizes
acceptable actions with a small amount of risk. Absolute perfection is very
expensive to accomplish, although high levels of quality may be strived for.
The increments of movement toward absolute goals, such as absolute zero
(temperature) or the speed of light (velocity), tend to get smaller while the
costs of approaching that absolute goal grow exponentially larger. It is the
same with safety—zero risk is extremely expensive compared with some
smaller finite value of risk considered acceptable.

A second common misconception in the field of safety is that its application is "just common sense," with the corollary that anyone can perform the
safety function because everyone has common sense. The problem with this

viewpoint is that safety engineers routinely need and use skills far beyond commonsense skills. Above all, scientific analysis must be used in the normal engineering context. Safety engineering is increasingly concerned with energy-exchange concepts and the safety of interfaces between systems, both not commonsense skills. Safety engineers depend upon the technologies being used in their discipline, and therefore must be conversant with emerging technologies. A strong case can be made for moving beyond a superficial safety concept that addresses only behavioral aspects. It is not enough to load up the human operator with most of the responsibility and consequences of avoiding losses, especially if we do not supply any information on how to control outcomes. I take a strong personal position against the primary use of the motivational approach to safety in that it is too vague, too ineffective over the long term, and too expensive. It is often a reactive posture, reacting to effects, rather than causes. However, just as absolute safety was deemed untenable, complete avoidance of motivational approaches is also unwise. In my estimation, motivational approaches should be used only after engineering approaches are in place. For this reason, motivational safety has been practically excluded from this text. In summary, there has been, there is, and there will be disagreement in industry over the definition and scope of safety engineering.

Timing

Another philosophical aspect that is often the subject of debate is that of when to perform the safety process. Just as an ounce of prevention is worth a pound of cure, before-the-fact safety is more valuable than after-the-fact safety. As will be decribed further in the System Safety section, a front-end emphasis allows earlier and smaller redesigns that normally cost less than waiting to react to losses by massive changes or a series of patches to hardware and operations. This is so because overhead resources may be used without having to bring them in specially, such as during construction. Also, the changes may cost less than the original design, and if done in place of the original, the costs are not additive. Changes made on paper still have an incremental cost, but it is small compared with redoing facilities and operational manuals. Retraining may also be avoided with front-end changes.

Change control and trade studies to select more optimal changes will be discussed later. Therefore, prospective safety is the preferred engineering approach to safety. Retrospective and reactive safety is more costly and disruptive to program organizations and their performance.

Finally, in nature, a normal state is one of minimal energy, if left alone. Thus, the normal state of safety within organizations will not be very safe, unless there are interventions. Without organizational direction, the need for safety engineering will be filled like a leak into vacuum, with whatever is adjacent, determining the quality and amount of safety. Organizational avoidance of making conscious choices about safety is equal to choosing mediocrity, or even worse.

2

SAFETY TERMINOLOGY

One of the marks of a profession is the existence of a unique terminology that has specific meanings. Safety engineering has extensive terminology, with well-defined meanings, but the basic word *safety* is often poorly defined and poorly understood because of lack of insight and, given the insight, lack of consistency. Furthermore, the term *engineering,* in a strict sense, may not be appropriate to describe what most safety engineers do. Thus definition of key terms is necessary to understanding of both content and scope of the safety engineering discipline.

SAFETY

Webster's Unabridged Dictionary gives seven explanations of safety, starting with *"the condition of being safe: freedom from exposure to danger."* Then in circular fashion, safe is defined as *"freed from harm, injury, or risk: no longer being threatened by danger or injury."* Alternately, the dictionary uses *"the quality or state of not presenting risks."* Further down the list of seven explanations of meanings, safety is described as *"knowledge or skill in methods of avoiding accident or disease."* In this case, the further you read more light is shed. Nonetheless, the general perception of safety is *"freedom from danger."* Yet this is what an engineer would call absolute safety. The difference between safety and absolute safety is similar to the difference between zero and absolute zero temperatures. In both cases, absolute is as low as you can go.

What the *ASSE Dictionary of Safety Terms* prefers to use for a safety definition is the more practical connotation of "relative safety." That is, safety is the risk state compared with what the organization is willing to accept.

9

- *Safety:* Relative acceptability of an operation in terms of losses or risks.

In other words, safety depends upon the organization's value system. Safety is a lot like beauty: it is in the eye of the beholder, the beholder being the organization. Of course, laws can legislate what standard of safety shall be used for measurement purposes. What the Occupational Safety and Health Act (OSHA) law demands is minimal acceptable safety based upon consensus standards. If companies fall below this minimum standard they are out of compliance. If they do much better than the minimum, there must be other reasons for safety than just compliance and, by the way, there are many compelling economic reasons for going beyond compliance.

OTHER RELATED TERMS

Accident is a close second to safety in being misunderstood. Webster's dictionary leads off with "an event or condition *occurring by chance* or arising from *unknown or remote causes.*" This element of mystery provides a convenient scapegoat, *implying that* fate is somehow involved and that *prevention is impractical.* This belief may be bolstered by a more legalistic definition: "an unexpected happening causing loss or injury which is not due to any fault or misconduct on the part of the person injured, yet entitles some kind of legal relief." Too often, the connotation of accident is "injury from uncontrollable cause or carelessness on the part of an individual." The safety professional needs a more pragmatic definition to work with.

The preferred definition of safety ignores the element of chance and broadens the undesired event to include anything which results in business losses. Note that the event is not restricted to that which happens on company property.

- *Accident:* An unplanned interruption of planned business activity, resulting in loss.

Examples of accidents include electrocution of a worker, collapse of a building, explosion of a pressure vessel, overturning of a tank truck, and release of a carcinogen.

Loss is a term common to the insurance business and connotes dollar loss. In fact, Dr. Ralph J. Vernon, retired professor of Industrial Engineering at Texas A&M University, writes it as "LO$$" for impact. Management tends to understand loss better when it is translated into dollars.

- *Loss:* Degradation of a system or component.

Examples of losses include death or injury to an employee, destruction or impairment of facilities or machines, ruining of raw materials, and creation of unavoidable delay.

Risk further refines the concept of a loss by including time and acceptability. Probability is used to relate loss occurrence over time. Acceptability of losses is an important attribute of risk.

- *Risk:* The result of a loss-probability occurrence and the acceptability of that loss.

Examples of risk include probability of one shuttle loss per fifty missions (cost on the order of $10 billion and seven lives lost, major delays in shuttle-supported space programs, and adverse political consequences), and the probability of a truck driver fatal collision per one year's operations.

System is a difficult term to define because of the great diversity in systems, varied interrelationships within systems, and because that system is what you want it to be and what you define it as.

- *System:* The whole set of factors, components, independent and dependent activities that interact in a predictable way to describe something within some limits.

Examples of systems include computer and printer; operator, computer and printer; aircraft radar; commercial cargo aircraft, crew, and ground support; and the U.S. Air Traffic Control Network.

System safety is a system concept applied to safety concerns.

- *System safety:* The application of systems engineering and management to the prevention and control of losses or risks for complex systems.

An expanded formal definition of system safety is given in the separate system safety section. Many other safety terms are defined in the sections in which they are used.

3

THE ENGINEERING APPROACH

THE SAFETY DICHOTOMY

Safety engineering may be thought of as having two separate points of application: one that separately evaluates the overall status of an operation and another that basically attempts to accomplish or improve organizational safety performance. In smaller organizations, the safety staff must wear both of these functional hats—the evaluator and the advisor. In larger organizations, there is a separation of these functions so that a naturally occurring accommodation and dilution is less likely to occur. It is similar to the checks-and-balance system in government—to have a "disinterested" party evaluate the safety engineering services to the organization. That is, those who design and do safety should not be the same ones who judge its adequacy. This chapter focuses on the doing function as opposed to the auditing function, although there are obvious overlaps.

Action vs. Problem Recognition

One greatly deserved criticism of safety engineering is that it often merely identifies problems and then does nothing to help with their solution. As true team players in the organization, safety managers and safety engineers must engineer solutions to safety problems. Importantly, they must engineer solutions to be optimal or they must at least recognize the effects upon organizational processes.

Strategy of Doing Safety

The Recognition-Evaluation-Control (R-E-C) concept is a fundamental engineering logic strategy. In several variations, it is the R-E-C process. Figure 3-1 illustrates the process.

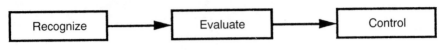

FIGURE 3-1 The R-E-C Process

Recognition

In safety engineering, recognition relates to the identification of hazards. A basic knowledge of hazards and how these hazards can lead to losses is needed by the analyst. Coupled with this knowledge, observation comes into play. Observation involves skill. The main objective of observation is to recognize exceptions to the acceptable norm. The norm yardstick must be scrutinized because what is normal may be unacceptable from a safety standpoint.

All senses may be involved in identifying hazards, but vision is primary. Even if the observer is concentrating on looking, the visual sense is only 83% efficient. Additionally, only 40% of the space around a person is in view without turning the head. Recognition may not occur unless there is sufficient time for observation. Some safety services companies offer safety program training that addresses and refines the recognition process.

Safety inspections are a common form of recognition. As with many comprehensive activities, checklists can improve the process by requiring less knowledge on the observer's part. A generic safety inspection observation checklist might include:

- Biological agent work (especially pathogens)
- Calendaring operations
- Chemical work (especially labs)
- Compressed gas operations
- Confined space work
- Construction work
- Electrical work
- Elevated work (including work on scaffolds and ladders)
- Hazardous materials work
- Hoisting/lifting (manual and mechanical)
- Hot work (flame and/or spark-producing work such as welding)
- Machine shop work
- Personnel-protection-required work

- Point-of-operation work (requiring guarding)
- Powered tool work
- Powered truck operations (including warehousing)
- Radiation operations
- Repetitive motion operations
- Solvent operations
- Spray-painting operations

Because many hazards are related to the exchange of energy, understanding of the following concept is needed for effective observation.

The Energy Exchange Concept

- Free energy transfer in any uncontrolled manner generates damage in people and machines.
- Energy forms include thermal, electrical, kinetic, potential, chemical, and radiation energies.
- Large reservoirs or high magnitudes of energy can generate drastic effects.

Safety engineers must be vigilant for energy exchange in their observations, which form the first part of the R-E-C process.

Evaluation

After hazard recognition is complete, the second step is hazard assessment or "evaluation," the "E" in the R-E-C concept. Hazard assessment requires both a risk determination and a risk judgment decision. Only after the risk decision is made should controls be undertaken. Where there are risk alternatives, a further step by management is involved, that of risk acceptance.

Mo Ayoub's temporal concept of hazard analysis follows. Past hazard analysis uses assessment of accident data. Present hazard analysis uses activity evaluation and predictive models. It is important to recognize that a hazard analysis is most powerful if applied prior to realization of any losses. But a continuum of hazard analyses needs to be used for the prevention of unacceptable losses—past, present, and future.

Where past performance is known, safety indices such as frequency, severity, and cost information may be applied to predictive or corrective studies. Other more specialized data presentations by attributes such as by temporal value (year, month, time of day), experience, and so on, are discussed later in the text. Control chart technology from quality control can be applied.

In addition to inspections and surveys, analyses based on current procedures, such as Job Safety Analysis (JSA) are used to evaluate present situations. High Potential (HIPO) and Critical Incident Techniques focus on a meaningful minority of the actions and causes, and are used in a predictive manner. Note that in this book, capitalization has been used to emphasize the names of techniques and subjects which form the basis for acronyms.

Predictive modeling, testing, and pilot plant operations are examples of mainly forward-looking hazard analysis techniques. In particular, computer-based modeling is increasingly being used to assist hazard analysis. Also, most system safety analyses are intended to be used in the predictive mode.

These are the precursors that must be satisfied prior to an analysis: (1) objective or objectives, (2) scope, and (3) depth. Data may need to be weighted to emphasize significant factors. The ability to weight data depends on an adequate understanding of the system being analyzed, influenced by the analyst's experience. Errors in weighting choices can be damaging to the results just as process logic errors can induce Single Point Failure (SPF) potential into the analysis. An SPF has the potential to fault the system without need for any other failures. For these reasons, the system must be well comprehended before proceeding with evaluation techniques.

In reality, one never quite knows everything to look at, or what to understand during the development of an evaluation. Experience can lend insight to repetitions. However, safety engineers would do well to remember Heizenberg's principle, which says that the mere process of observation will affect the outcome. The eye of the beholder decides what is beautiful, people respond positively to attention (commonly referred to as the Hawthorne effect) and expectations tend to influence performance. Analysis is a tricky task.

Control

Once recognition has led to evaluation, the last phase—control, can take place. Although some elements of organizational politics may be present in recognition and evaluation, politics is certainly involved in the decision making and implementation associated with control. A good, systematic job of recognition and evaluation can provide a less politicized base on which management may take its control actions.

Control involves the detection and correction of adverse change so that a system performs as intended. Control can be done before you do something (at least before you set it into motion), while you are doing something, and after you do something (so that you can do it better next time). Control specialists call this feed-forward or pre-process control, in-process control, and feed-back or post-process control, respectively.

From a safety engineer's viewpoint, control of hazards keeps the potential from becoming real. Control can also be referred to as an "inhibit." That is, it inhibits the undesired action. Inhibits can be physical barriers, such as a shunt to carry current to ground, or the lack of a connection that must be inserted to connect the circuit. Or inhibits may be procedural or software actions required to "arm" the hazard into a dangerous state. Individual controls are picked according to a control precedence.

A hierarchy of safety controls (also called the hazard reduction precedence) follows:

1. Eliminate the hazard (do not do something or do it a safer way—a methods improvement).
2. Overdesign (design to minimum risk).
3. Control by failsafe design and fault tolerance techniques.
4. Contain the hazard (limit the damage possible—barriers, isolation).
5. Harden the system against damage.
6. Provide safety devices (pressure relief valve, personal protective equipment, and so on).
7. Train personnel to adapt by procedures.
8. Caution, Warning, and Alarms (take corrective action, flee the hazardous area and/or take precautions).
9. Do nothing (justify residual risk).

Multiple controls are better than single controls; controls can be of like design or of different design. Multiple controls of different design are generally better than like design because of a common cause's ability to defeat all redundant controls.

4
SAFETY SKILLS

Safety skills involve knowledge and proficiency topics in areas that are common in safety engineering. The safety skills listed in the following sections are considered useful for the majority of safety engineers. Safety skills are numerous and difficult to categorize without overlap. Some safety skill areas, such as nuclear safety engineering, have been omitted because these areas are highly specialized and are beyond the scope of this text. Safety skills is a term of convenience and includes industrial activities, common causal factors and, in some cases, major areas of effect, plus some disciplines that control hazards and risks according to a particular set of concerns. Readers of this text should realize that the following collection of safety skills are not homogenous. I decided to treat them as a whole, rather than further subdivide the subject, as would be necessary to group all effects separately from causal factors or industrial activities.

In this section, various safety skills are discussed in terms of their importance to industry, such as the magnitude of the problem, typical causes and other important factors, and typical controls; where possible, lists and sources of additional information are provided. There is an attempt to use the more readily available references, such as the American Society of Safety Engineers magazine *Professional Safety*. There is also an attempt to use more recent references, although occasionally better information is given in an older article or book.

Readers should recognize that the information explosion applies to safety, and therefore I may not be aware of many additional good information sources which have been published. This book was written using collections of information gathered since the mid-seventies. Other than normal professional reading, an information search was not done as a part of writing this book.

Facts and figures used as illustrations in this text should be considered indicators more than benchmarks upon which to make decisions. Invariably, figures depend on interpretations and ground rules. Readers are warned to continuously seek updated information, and collect safety information on their own to update this text.

There are many inferences, suggestions and, no doubt, conclusions with which you may not agree. This is to be expected. Safety engineers, like anyone else, should not be timid about searching for their own truths and should critically evaluate everything read and heard.

Safety skill areas are listed alphabetically for convenience because it would be difficult to arrange them by importance.

ACCIDENT INVESTIGATION

Accident investigation is corrective action taken after the fact, aimed at identifying the facts and underlying causes of accidents. Since the word "accident" is subject to various definitions, the process varies widely. Using our earlier definition about unwanted interruptions to business activity, accident investigators need to look into loss events and near-loss events.

Ludwig Benner, Jr. lists five perceptions of the nature of accident phenomenon that can influence the theoretical bases for accident investigations (Benner, 1982). They are:

- *Single-event perception:* primitive, simple explanation of a phenomenon in terms of one event, one cause, one "scapegoat."
- *Chain-of-events perception:* formalized by Heinrich in his "domino" theory in which unsafe conditions are toppled by an unsafe act. Cause is modified by such terms as proximate, primary, remote, and so on, to accommodate complexity.
- *Determinant variable or factorial perception:* search for the ideal single independent variable and reliance upon examining facts to isolate factors not due to chance.
- *Logic tree perception:* assumes a converging chain of events leading to an undesired event. A logical presentation allowing event–condition relationships to be viewed simultaneously and predictions of effects when event probabilities are known.
- *Multilinear Events Sequence (MES) perception:* an accident process described in specific interacting actors, each acting in a sequential order with discrete temporal and spatial logical relationships.

Benner additionally describes strengths and weaknesses for each of the five perceptions and finds logic trees and MES most promising approaches for accident investigation.

Accident investigation should focus on causes and events, not individuals. Blame is not the objective. The fixing of blame should be left to strictly legal and judgmental forums. It is important to determine valid, appropriate facts, just as in a legal determination, because the determination of cause depends on good fact-finding. A properly done accident investigation will provide a good deal of common information that is also needed in the accountability function. The adage "If it wasn't documented, it didn't happen" applies to accident investigation and accountability actions.

Accident Investigation Plans

Just as with surveys and other studies, it is necessary to decide beforehand what must be investigated, how to go about the investigation, what resources are needed, and what the output of the investigation should look like (not the results, but the format). Unplanned investigations rarely yield valuable results.

An important point is that the highest potential for losses should warrant the greatest amount of effort and resources in terms of accident investigation. Exact delineations of what, how soon, and how to investigate are management decisions, just so long as management recognizes what situations can cost their organization the most.

Obtaining written statements from witnesses should be preplanned and be clearly voluntary on the part of those interviewed. Similarly, any voice or video recordings must be agreed to in advance of interviews. If the gaining of individual permission is not clearly stated in the plan, it may not be received, and may cause legal problems.

Evaluation and analysis methodologies must be spelled out in an accident-investigation plan. A tailored software program is a good way to assist team members in the process, simplify data acquisition and capture, and facilitate analysis. Use of a notebook-type computer, of sufficient power, can allow on-site entry of data.

As organizations should have written safety plans, they also should have written accident investigation plans. Such formalized planning not only helps the quality of the plan design, but it also allows management to project manpower, equipment, and interface requirements. Accident investigation is a bona fide, learnable skill and preparation is therefore necessary. This skill is important enough to the organization that it should be backed up by written guidelines so that others within or outside the organization can perform the investigation, in case the principle investigator(s) is unavailable.

An accident-investigation manual should be prepared for each organization, reflecting predefined investigation procedures, responsible team members and boards, required documentation, and the necessary interfacing with
company public relations, legal, and personnel specialists.

Investigation Personnel and Methodology

Accident investigation rarely operates in a vacuum, so it may be helpful for
the investigator also to be a skilled team leader. Someone who is well
grounded in the system will be needed. Investigator and board member
training is necessary in advance of accidents. Similarly, the tools of accident
investigation need to be identified and set aside for rapid response. Multiple
sets of tools may be needed if the risk of equipment loss is high enough, if
there could be multiple investigations simultaneously, or if the investigations
occur at a distance. Many times, investigators must throw a kit into an
airplane or a vehicle and race the clock to respond before the information has
evaporated. The circumstances of the operations dictate the people to be
involved, the equipment, and the interfaces necessary to accomplish accident
investigation. It would be fruitless to attempt detailed contents of a plan or
an investigation kit here without knowledge of the situation. However, as an
example, an investigation kit should include

- Bound notebook and pencils
- Camera (35mm, black & white film (ASA 400); Polaroid color camera
  with flash)
- Folding carpenter's ruler
- 100′ surveyor's tape
- Mini tape recorder, batteries, tapes
- Sealable plastic bags, tags, indelible marker
- Plastic barrier tape ("Keep out—investigation site")
- Phone list of key contacts
- Programmed (for accident investigation) laptop computer
- Video camera with low light and sound capability

Larger organizations may organize their accident investigators, equipment,
and transportation by teams and functional areas throughout the business and
may even have an accident investigation manager. Other organizations may
be expected to incorporate accident-investigation functions, especially the
more proactive kind, into safety- or loss-control organizations.

The timing of accident-investigation response can generally be categorized this way: "as quickly as it can be done safely." The investigation should be done concurrently or on the heels of the emergency response. A lot of activity that must logically take place in response to an emergency is counter to the accident investigation, if facts are not recorded somehow.

Photography, sketches, statements, and reports are very helpful to the investigation. OSHA's Jennifer Miller urges investigators to collect the four P's: **P**eople interviews and statements, **P**osition measurements and photos at the scene, **P**arts such as chemicals and broken pieces, and **P**aper forms, records, and the like (LaBar, 1990). Photography and interviewing skills are valuable. Leading the witnesses tends to affect the investigation outcomes adversely.

Checklists of facts to gather can assist the investigator. Ted Ferry has detailed an excellent generic approach to mishap investigation, including a twelve-step mishap investigation process ([Ferry, 1981], reprinted with permission):

1. Understand the need (be prepared phase)
2. Prepare for the investigation
3. Gather the facts (information development phase)
4. Analyze the facts
5. Develop conclusions
6. Analyze conclusions
7. Make the report (action phase)
8. Make appropriate recommendations
9. Seek corrective actions
10. Follow through on recommendations
11. Critique investigation
12. Double check the corrective action

Proceeding from cause to effects is an art, closely allied with problem solving. Problem-solving techniques such as those of Kepner and Tregoe should be applied to identify causes. It is important to remember that multiple causes are more often involved than just one or two causes. Some method of considering multiple causes must be used. For instance, in an Australian train accident that took eighty-three lives, fifty essential contributing factors were identified (Emerson, 1985). The distinction between a "contributing factor" and a "basic cause" requires discernment. Virgil Casini of the National Institute for Occupational Safety and Health (NIOSH) suggests a nearly endless list of sometimes overlapping causes (but points to compliance with existing standards as the cure to most workplace deaths [Emerson, 1985]).

Keep in mind that the end result of accident investigation must include a cause–effect determination. Although there are many ways of analyzing for cause and contributing effects, I believe the following techniques to be excellent for accident investigation. Change analysis focuses on differences between normal, controlled operations and the accident. Management Risk & Oversight Tree (MORT) looks for management inadequacies as causes. Fault Tree Analysis (FTA) can be used retrospectively to determine accident sequences and causes. Linear Event Sequencing and Digraphing techniques may provide additional insight as to cause and effect. Allison's High Potential (HIPO) analysis differentiates potentially serious loss situations from less severe incidents and has use in recommending corrective actions. Details of the previous analyses are available elsewhere in numerous safety engineering and system safety texts.

A Los Alamos change-based accident analysis worksheet (Bacastow, 1978) is given in Figure 4-1.

Samples of MORT questions regarding accident investigation (Johnson, 1984) are as follows:

Fundamental Questions

- What happened?
- Why?
- What were the losses?

Barriers Less than Adequate

- Were there adequate barriers on the unwanted energy?

Barrier Failed

- Did the barrier prevent transfer of energy as designed?

Did Not Use

- Were barriers used?
  a) Did not provide
- Were barriers provided where possible?
  b) Task Performance Error
- Were provided barriers used properly?

Inspection Less than Adequate

- Was there adequate inspection of equipment, processes, utilities, operations, etc.?

Factors	Present Situations	Prior Comparable?	Differences?	Affective changes?
What: Objects, Energy, Defects, Protective Devices				
Where: On the Object, In the Process, Place				
When: In Time, In the Process				
Who: Operator, Fellow Worker, Supervisor, Others				
Task: Goal, Procedure, Quality				
Working Conditions: Environmental, Overtime, Schedule, Delays				
Trigger Event				
Managerial Controls: Control Chain, Hazard Analysis, Monitoring, Risk Review				

Source: Los Alamos Scientific Laboratory LA-7877-MS.

FIGURE 4-1   Change-based Accident Analysis Worksheet

- Was the plan or its execution less than adequate?
- Are inspectors trained in hazard recognition?

Job Hazard Analysis

- Had the job been analyzed?
- Did the analysis identify the hazard and appropriate safe work procedure?

Training

- Was the employee adequately trained in safe work procedures and hazard identification?
- Was the supervisor adequately trained?
- Were inspectors properly trained?

Human Factors Engineering

- Was consideration given in design, planning, and procedures to human characteristics as they interface with machine and environmental characteristics?

Employee Selection

- Were methods of personnel selection and placement adequate?
- Were the safety-related job requirements adequately defined so as to select and place the individual with adequate qualifications?

Health Hazard Controls

- Were people and objects free from physical stresses caused by health hazards?

Accident Investigation

- Was the accident investigation program less than adequate?
- Have similar accidents occurred which were not properly analyzed for cause and appropriate corrective action?

Emergency Plans

- Was there adequate emergency action to prevent a second accident from occurring?
- Was the plan executed adequately?
- Was the plan designed adequately?

OSHA Compliance

- Were there written procedures to assure compliance with applicable OSHA requirements?

Final Documentation

The following suggested arrangement for a final accident investigation report is adapted from Bacastow (1978):

a. *Transmittal Letter* (authority, signatures)

b. *Cover* (subject, date of occurrence, date of report)

c. *Table of Contents* (sections, subsections, illustrations, charts, appendixes)

d. *Scope* (issues, objectives, area, instructions, and limitations)

e. *Summary* (overview of essential facts, findings, conclusions, and recommendations contained in the report)

f. *Facts* (logical development of facts)

g. *Analysis* (description of organization, interpretation, sequences, inferences)

h. *Conclusions* (sequences, analytical findings, probable causes, dissenting positions)

i. *Recommendations*

j. *Appendices* (diagrams, flow charts, exhibits)

In the final report, documentation should follow a plan-directed collection. A final report should contain data, analysis methods, conclusions and recommendations, and be signed by responsible team and board members. Photographs and simplified flows or schematics should adequately describe the process or situation. Physical evidence should be carefully described along with the statements acquired from individuals. Revealing of names in reports should be cleared with attorneys before release. Detailed and lengthy support information should be relegated to attachments to the main report.

AGRICULTURAL SAFETY

Agriculture is the second most dangerous large industry in the United States; only mining and quarrying kill more people, according to the National Safety Council (NSC) (Hair, 1991). But compared with other industries, agricultural safety is less regulated than other fields. Many farms are small operations with fewer than ten employees. Nevertheless, agriculture has a great many hazards that result in many deaths and serious injuries each year. Jack Burke stated in the *American Society of Agricultural Engineers Journal*, "Agriculture is the largest, oldest, most fundamentally important of the nation's industries. Yet it is the only major industry without a safety program of national scope" (Hair, 1991).

Nationwide statistics in agricultural safety are not as comprehensive as for other industries. NSC Accident Facts reported about 2,800 farm deaths (58.6 per 100,000 population) and more than 250,000 disabling farm injuries in 1988. Tractor rollovers (3.7 per 100,000 tractors) result in 44% of all agricultural deaths and many of the serious injuries. The 5–14-year-old age farm population group suffered the highest injury rate at 27.2 per 100,000 population. Rates of death and injury were higher in hired workers than in family agricultural workers (National Safety Council, 1989).

Hazards

Hair lists four areas of farm safety hazards: farm machinery, occupational health, ergonomics, and animal hazards (Hair, 1991).

Specific significant hazards include overturns, confined spaces associated with commodity storage, rotating equipment, crushing/cutting hazards from power takeoffs, rotating mower blades, rotating augers, and rotating conveyor belts common to many types of farm equipment. Much of the equipment is aging and without effective guarding of hazardous points of operation.

Tractor rollover often occurs as a normal consequence of tractors pulling loads up inclines with strong torque forces causing rotation about the rear axle. Tipovers result from exceeding the friction forces due to sloping terrain and high centers of gravity. Rollover protection systems (ROPS) are present only on 20% of tractors, and the companion feature of safety belts are used infrequently, defeating part of the protection of the rollover cage (Hair, 1991).

In-running nips and shear points exist with many power takeoff arrangements. Lack of guarding causes many entrapments of extremities or clothing, which result in deaths or serious injuries due to the high power of the rotating shafts. Other rotating equipment is contacted when farm personnel fall from moving vehicles and are struck by mower blades. Another common scenario is when workers are caught by rotating equipment when attempting to clear jams with the equipment running.

Many farm machinery operations are noisy and farmers suffer commensurate hearing loss. A Wisconsin study showed higher loss percentages with farming than with nonfarming populations, with workers demonstrating loss by age 30 (Hair, 1991).

Confined spaces exist in silos and grain bins. Hazardous atmospheres may be generated by natural processes or by the introduction of inert atmospheres to control animal and insect vectors. For instance, workers have been killed or seriously injured when unloading products from railcars lined with phosphine-producing pesticide tapes.

Pesticide applications have poisoned many workers, often as a result of mechanized power spraying or aerial cropdusting. Some of the more persistent and high-toxicity pesticides have been removed from agricultural use, but still many products require professional-application safety precautions.

Herbicides and growth stimulants are less of a toxicity problem but have potential chronic effects.

Fungi spores and certain bacteria may be injurious to some agricultural workers. Infectious disease may arise from agricultural products or from animals.

Anhydrous ammonia, used as a liquid fertilizer, is a hazardous lung irritant that has caused many deaths. This highly caustic material also readily damages the eyes and unprotected skin.

With the advent of piped irrigation, many workers are electrocuted either by contacting power lines with conductive pipes or when becoming a part of the ground path of improperly grounded electrical motors and pumps.

Still other workers are killed in falls from elevations, while harvesting fruit. The pollination of dates often involves working on very high ladders or long-reach positioning aids such as hydraulically operated cranes (incidentally known by their agriculturally inspired name—"cherry pickers.")

Farm animals are capable of inflicting serious injury or death, especially large animals. Direct handling of animals in chutes or stalls may lead to crushing or goring potentials. And the administration of medicine can provoke otherwise docile animals.

Agricultural injury studies in California show that musculoskeletal injuries are most common, due to the large amount of manual labor involved in agriculture (Hair, 1991). Back injuries may be due, in part, to driving for many hours in poorly designed seats, exposed to whole-body vibrations. Many agricultural working surfaces are slippery or uneven, leading to slips and falls.

Cuts are often aggravated by dirt and bacteria associated with agriculture. Injuries may occur at long distances from first aid equipment or providers.

Controls

Many hazard controls from other types of industry can be applied to agriculture. Torque-indicating and torque-limiting devices can reduce rollovers. Inclinometers can help prevent tipovers. ROPS and seatbelt use can keep workers inside a safe envelope if tipovers occur. Safe operating procedures should be taught. Subpart W of OSHA CFR 1926 gives ROPS performance criteria and testing information for wheel-type agricultural tractors used in construction. The tests, especially the turnover tests, should be of value.

Standard confined-space entry procedures can be applied to silos and pits. Calibrated testing for oxygen deficiency and toxic gases should be adopted. Protective equipment such as self-contained breathing apparatus and rescue harnesses should be employed.

Proper respiratory and skin-protective equipment should be used when applying pesticides, herbicides, and fertilizers, and with biohazardous agricultural products. Reentry times should be adhered to. Biological testing for overexposure to toxic materials should be used whenever appropriate.

Hazardous energy sources should be physically guarded wherever possible. Lockouts should be used with larger pieces of equipment and physical protection such as interlocks, and blocks should be applied to equipment to avoid inadvertent start-up hazards. Jam-clearing procedures should require operators to shut off the machinery engine or power source.

Electrocution can be avoided by good electrical protective design and practices, and by using two workers to carry conductive irrigation pipe rather than one vertically balancing a long section while walking.

Fall protection should be used to protect workers at elevation. Seatbelts should be used to keep operators safely in place on mowers and other equipment without closed cabs. Dead-man controls should be used for machinery that can injure an operator out of the normal operating location.

Large-animal handling should be done with appropriate safeguards and equipment such as hold-off devices, squeeze chutes, prods, and more-than-one-person operations.

Manual handling strains can be reduced by task redesign, use of mechanical lifting aids, or multiple lifters. Vibration hazards may be lessened by proper maintenance and incorporation of vibration isolators. Seats should be lined with foam.

BACK INJURY

Effects

Because of the magnitude of the back injury problems in industry, it may be appropriate first to focus on the effects and then on the underlying causes. As has been stated before, it is futile to try to control an effect, rather than the cause.

Back injury, as a multicausal effect, is the largest single injury cost within the safety arena. For our purposes, back injury includes low-back pain, although technically, injury may be restricted to strain, sprain, and fracture. Many researchers have estimated that, over time, 60–80% of the working population will experience significant low-back pain and 90% of these will have at least one recurrence (La Bar, 1991; Snook, 1980). Backache represents 40% of all recorded absence from the workplace, second only to the upper respiratory ailments of the common cold (Owen, 1986). White's study of 12,000 low-back-pain cases showed that all but 20% of workers were able to return to work within three weeks (Snook et al., 1978). However, of the 2% of Canadian workers off work for more than one year due to back injuries, 75% never return to full-time employment. The U.S. situation is similar, so it is no surprise that only 20–25% of all back claims account for a significant 90% of costs (Pareto's critical few principle) (Maravino, 1989). Overexertion accounts for about one-third of all occupational injury and illness cost, and 68% of these are associated with lifting. In 1985, average costs per incident were about $3,000 (Liles and Mahajan, 1985).

Causes

For many years there has been controversy about what proportion of back injuries should be attributed to work-related causes and how much should be ascribed to normal degeneration, off-the-job causes, and even fraud.

The safety engineer should recognize that job-related back injuries can have their roots in (1) the worker's performing work, (2) the task design, or (3) a combination of the two. Too often in the past, the worker received the major emphasis, if not blame, for the resulting injuries. I believe that the greater fault often lies in the task design, considering not only the ergonomics but such things as an incentive system that drives the work rate and willingness to assume additional personal risk.

M. Laurens Rowe's long-term study of 500 industrial back injuries showed that about 65% of back injuries occurred while employees were doing normal work (Sorock, 1981). Snook states that lifting is indicated as a cause in up to half the manual material-handling task injuries, while pushing, pulling, and carrying roughly make up the remainder. Importantly, twisting and bending while lifting or carrying has been related to onset in about one-third of back pain injury (Snook, 1988).

Lifting and Ergonomics Research

Many ergonomics studies have improved our knowledge of the forces involved, especially in manual material-handling (MMH) tasks—in fact, too many to credit here. There have been static studies, dynamic studies, subjective studies, and performance-related studies across all sorts of populations—measuring weights, symmetry, heights, and repetition rates, with many attempts to pin certain variables. Needless to say, not all studies were of the same scientific caliber, and some results conflict. While a comprehensive treatment of this subject is impractical here, in general, men lift somewhat more than women, but both sexes reach reasonable lifting limits that can be described in statistical parameters such as age, type of task, repetition rate, and physical hardening. The ability to relate the physiological demands to injury and illness are more variable, probably due to individual etiology.

The ergonomics and work physiology criteria have many different forms, such as lifting–strength rating, job severity index, and psychophysical tables, but the NIOSH Work Practices Guide for Manual Lifting guidelines seem to have made the most impact, despite some limitations (Snook, 1988). Many models have been developed for biomechanical and metabolic lifting, using static, isometric, isoinertial, isokinetic, isodynamic, and psychophysical measures of strength. OSHA indicates that although ergonomics is one of the major occupational safety and health issues of the 1990s, a general industry standard may not be in force until the end of the decade or later (*Occupational Hazards*, 1991).

Recently many of the long-cherished maxims about lifting techniques have come under scrutiny, based on new biomechanics findings. In particular, the straight-back, bent-knees lift method may cause increased loads at vulnerable L5/S1 lumbrosacral disc and demand muscle strength beyond the capability of many lifters. A computerized biomechanical analysis of lifting a 100 lb bag from a single pallet height above ground created a 1500 lb L5/S1 compressive force, which relates to an eightfold risk level of low back pain (Anderson and Catterall, 1987). The strongest criticism should be the lack of improvement in back injury rates despite the lengthy emphasis upon the straight-back lifting method. David Apts' American Back School advocates an arched-back lifting technique (Lawn, 1987). Don Chaffin states that about one-third less compressive stress is placed on the lower back than the squatting type of lift (Sorock, 1981). Better advice includes the following generalities: (1) Keep the load as close to the body as is practical, (2) Avoid rapid jerking of the load, (3) Avoid twisting or bending with the load during lifting, (4) Supply loads between knuckle and shoulder height, (5) Avoid lifting above shoulder height, (6) Control the pace of lifting, (7) Control the time to hold weight, (8) Anticipate the need for unexpected movements, and (9) Use a buddy system or mechanical devices for heavy and bulky objects (Owen, 1986).

Much of the basic research about MMH stress (task as the cause) can also be related to a focus on the individual as a causal factor. In the ergonomic attempt to better match people to the job, technology has been developed to evaluate strength testing. The incorporation of computer technology into strength testing has simplified the evaluation and repeatability.

With the trend toward more females, older workers, and better job access for handicapped workers, this matching becomes both more difficult and more needed. We need to establish a much better baseline regarding job task requirements, especially regarding muscular strength, dexterity, and so on. The

U.S. Air Force has developed a criterion for assigning personnel to Air Force Specialty Codes (AFSCs). (The U.S. Army also has a similar codes called Mission Occupational Specialties [MOSs].) Twenty-five tasks were used to represent physically demanding tasks for all AFSCs. Correlations were obtained from supervisors and task performers as to strength and endurance requirements, followed by base visits for real-world verifications. Using statistics methods, personnel and tasks could be matched (Ayoub et al., 1987).

James Ashton-Miller and Don Chaffin, University of Michigan researchers, note that older workers need not be automatically barred from jobs with heavy lifting. Heavy lifting is associated with an eightfold increase in the rate of medically treated low-back pain. But the same load can cause levels of strain differing by a factor of ten in individuals of the same age and gender. No study indicates that the frequency and severity of low-back pain is greater in older workers than in younger workers performing the same task. The larger risk factor is the frequency of physical activity. Prediction of functional capacity based upon age alone is fruitless. Medical supervision however, of older workers is important (*Professional Safety*, 1991). The U.S. Council on the Aging advocates use of the Koyl's GULHEMP (**G**eneral physique, **U**pper extremities, **L**ower extremities, **H**earing, **E**yesight, **M**entality, and **P**ersonality) job-rating program for placing older workers on the job (Snook, 1988).

Back Injury Controls

The primary controls of back injuries include elimination of high-stress manual materials-handling tasks and replacement with machine-assisted methods, manual design/redesign of manual lifting tasks such as by limiting weight (per NIOSH Work Practices Guides or other proven regimes), optimizing the lifting heights, distances, and orientations; improving handling aids such as handles and grips; use of multiple persons for heavy and/or bulky objects, and protective equipment. Secondary controls include selection of the most physically capable individuals for demanding tasks and preplacement exams, physical fitness evaluation and training, task-oriented training, and medical monitoring. Tertiary controls include such things as warnings and signs, behavioral appeals, injury rehabilitation, and transfer to less demanding jobs after injury.

The use of lower-back X-Rays to screen for predisposition to back injury does not seem warranted in that the radiation risk exceeds the benefits of identifying a small population with risky back deformities that could not be isolated by other low-risk methods. A NIOSH-sponsored 1973 nationwide conference of experts concluded that roentograms had no medical predictive value (Rowe, 1982). Dr. Rowe estimated that careful history taking and thorough examination could identify, at the most, 10% of young individuals prone to future back problems (Snook, 1988). It is possible that another medical test, thermal imaging, could assist in qualification of recent back injuries.

Some of the devices used to treat back injuries include posture correction seats, vest, and belt back supports. In my opinion, the preventative aspect of such devices is conjectural.

Treatment

Malcolm Brahms estimates that only 10% of industrial back injuries will require sophisticated medical testing, and that 90% of these represent muscle spasms and direct traumas that produce only temporary disabilities (Nemec, 1990). Only 1–4% of back injury patients have clearly identifiable and surgically correctable problems (DiMaggio and Mooney, 1987). The great majority of back pain patients get better, whether they seek medical attention or not. Two days of rest (rest plus drug therapy is the "standard" care) was found to be as effective as seven days of rest. Other methods of treatment include various methods of traction, bracing of ambulatory patients (less effective than autotraction), manipulation, and exercise. A large British study of chiropractic care indicated that 7 visits over 4.5 weeks on the average were required to achieve maximum results, with about one-third of patients needing to have continuing follow-up manipulations (DiMaggio and Mooney, 1987). The McKenzie program of flexion and extension exercises, combined with lateral bending and rotation, reduce peripheral symptoms to the point where posture changes and continued exercise can eliminate back pain. The McKenzie program has been judged superior to 90/90 traction and back schools advocating flexion exercises (DiMaggio and Mooney, 1987). The Williams brace and exercises are questionable as to benefit for severe disc pain patients from spinal architecture and a one-to-one correspondence with effects (Snook, 1988).

"Back Schools," many of which combine education with training, claim upwards of 70% reduction in back disabilities, suffer from the lack of scientific justification using control groups. *Occupational Health & Safety* magazine's "Focus" section lists information on about twenty-five back-rehabilitation and back-injury-prevention programs. Keyserling et al. (1980), reported that neither employee training programs in safe-lifting techniques nor traditional medical screening programs (based on medical histories or low-back X-Rays) have resulted in any reduction in low-back injuries. The survey concluded that the most effective method of controlling injuries would be to design jobs to fit the worker, thereby reducing overexertion injuries by as much as 67%. The American Industrial Hygiene Association *AIHA Ergonomics Guide* isometric strength testing protocol has had over 3,000 workers safely tested without any reported injuries.

Snook makes the important point that equally important to the education of workers is the education of management, unions, and practitioners who treat back pain (Snook, 1988).

Ergonomic controls by task redesign are applicable to the control of back injuries. Benson published the following helpful list (Benson, 1987):

Principles of Task Redesign

I. Minimize Significant Body Motions
   A. Reduce Bending Motions
      - Using lift tables, work dispensers and similar mechanical aids
      - Raising work level to an appropriate height
      - Lowering the worker
      - Providing all material at work level
   B. Reduce Twisting Motions: Eliminate the need to twist by:
      - Providing all materials and tools in front of the worker
      - Using conveyors, chutes, slides, or turntables to change direction of material flow
      - Providing adjustable swivel chairs for seated workers
      - Providing sufficient work space for the whole body to turn
      - Improving layout of work area
   C. Eliminate the need to reach by:
      - Providing tools and machine controls close to the worker, to eliminate horizontal reaches over 16 inches
      - Placing materials, workplaces, and other heavy objects as near the worker as possible
      - Reducing the size of the object being handled
      - Allowing the object to be kept close to the body
II. Reduce Object Weights/Forces
   A. Reduce Lifting and Lowering Forces
      1. Eliminate the need to lift or lower manually by:
         - Using lift tables, lift trucks, cranes, hoists, balancers, drum and barrel dumpers, work dispensers, elevating conveyors, and similar mechanical aids
      2. Reduce the weight of the object by:
         - Reducing the size of the object (specify size to suppliers)
         - Reducing the capacity of containers
         - Reducing the weight of the container itself

- Reducing the load in the container (administrative control)
- Reducing the number of objects lifted or lowered at one time (administrative control)
  3. Increase the weight of the object so that it must be handled mechanically:
     - Changing the shape of the object
     - Providing grips or handles
     - Providing better access to the object
     - Improving layout of work areas
B. Reduce Pushing and Pulling Forces
  1. Eliminate the need to push or pull by:
     - Using powered conveyors
     - Using powered trucks
     - Using slides and chutes
  2. Reduce the required force by:
     - Reducing the weight of the load
     - Using non-powered conveyors, air bearings, ball caster tables, monorails, and similar aids
     - Using four-wheel hand trucks and dollies with large diameter casters and good bearings
     - Providing good maintenance of floor surfaces, hand trucks, etc.
     - Treating surfaces to reduce friction
     - Using air cylinder pushers or pullers
  3. Reduce the distance of the push or pull by:
     - Improving layout of work areas
     - Relocating production or storage areas
C. Reduce Carrying Forces
  1. Eliminate the need to carry by converting to pushing or pulling by:
     - Using conveyors, air bearings, ball caster tables, monorails, slides, chutes, and similar aids
     - Using lift trucks, two-wheel hand trucks, four-wheel hand trucks, dollies, and similar aids
  2. Reduce the weight of the object by:
     - Reducing the size of the object (specify size to supplier)
     - Reducing the capacity of containers
     - Reducing the weight of the container itself
     - Reducing the load in the container (administrative control)
     - Reducing the number of objects lifted or lowered at one time (administrative control)

3. Reduce the distance by:
   - Improving layout of work area
   - Relocating production or storage areas

Design vs. Behavior

In summary, of the three classic methods to prevent back injury, the first two (selection and training) are ineffective. Regarding preventional motivation or "awareness" approaches, Leonard Ring (1989) notes that awareness does not always translate into performance, as shown by costly smoking and drunk-driving awareness campaigns. Only the third method, an ergonomic approach of designing the job to fit the worker (such as designing to 75% of the population able to lift without overexertion), is an effective control, and then it is only partially effective (67%). Thus, 33% of injuries would occur anyhow (Snook et al., 1978). The NIOSH *Workplaces Guide to Manual Lifting* states that about one-third of the U.S. work force is required to exert significant strength in their jobs. Also, overexertion injuries occur at the rate of about 1 in 200 workers per year (Chaffin, 1987). There will be a remnant of back injuries, even if the best ergonomic approaches are used, but it will be reduced significantly. As better manual materials-handling guidelines become available, they should be applied.

CATASTROPHES

Catastrophic events are of interest to safety engineers because they represent a highly visible and concentrated concern. Catastrophic effects involve high severities and they trigger reactions due to large numbers of deaths and/or serious injuries, enormous costs, or loss of major resources. A partial list of fairly recent industrial catastrophes includes Flixborough, England (1974); Bhopal, India (1984); Mexico City, Mexico (1985); Pasadena, Texas (1989) (Minter, 1990); and Channelview, Texas (1990) (Minter, 1990). Catastrophes evoke strong fears and reactions from the public and legislators. Prevention and control of these effects should be a major focus of safety engineering, in that by controlling catastrophes, lesser effects will be controlled as well. Certainly, the control of catastrophes must be based on control of causes, rather than the effect itself.

Webster's Seventh New Collegiate Dictionary (1971) defines "catastrophe" as a momentous tragic event ranging from extreme misfortune to utter overthrow or ruin. For this text, catastrophe is taken to mean multiple injuries, loss of lives, and major loss of property and business functions. Insurers usually report catastrophic loss in excess of many millions of dollars (including litigative costs and awards). Environmental disasters will be treated elsewhere, although loss of life is certainly possible from chronic effects of those catastrophes.

Occurrences

Explosions represent a major portion of the catastrophes, along with toxic gas releases like that in Bhopal, India, with its 3,400 deaths and 200,000 injuries. Processes that operate at extreme pressures and/or temperatures, with large amounts of energy being contained are prime candidates for catastrophe. Add to this flammability and toxicity considerations, and the problem is even worse. Cryogenic gas liquids that expand by orders of magnitude, superheated liquids under pressure, pyrophoric, and highly reactive materials are worthy of carefully engineered controls. Several catastrophes have been associated with the boiling liquid expanding vapor explosion (BLEVE), and vapor cloud explosions (VCEs). Common catastrophic toxic materials include ammonia, chlorine, cyclohexane, ethylene, hydrogen sulfide, liquified petroleum gases and liquified natural gas, methyl isocyanate, vinyl chloride, and highly radioactive materials. Explosive substances include intended explosives and propellants, but also explosive dusts such as in grain elevators and prilled ammonium nitrate fertilizer.

Steven K. Hall describes gas and liquid Categories I through IV, which are based upon critical temperatures. He also discusses common mechanisms for loss of containment (Hall, 1988).

The potential for major catastrophe which from transportation and fires in high-rise corporate office buildings is covered in the transportation safety, office safety, and fire safety sections. Underground mines, especially coal mines, have suffered catastrophic explosions over the course of many years. Mining safety is also covered in a later section.

Not all catastrophes involve fires, explosions, and highly toxic chemicals. The structural collapse and capsizing of a North Sea semisubmersible oil rig killed 123 people in 1980 and triggered more than $2 billion in lawsuits (Densmore, 1982).

Response

OSHA responds to public pressure about major loss of life with requirements for rapid notification and sends investigators for all loss-of-life reports. Catastrophes invariably command an accident investigation team. The recent process safety standard resulted from concern about catastrophes.

Governments tend to react to catastrophes with punishment and regulation. The public reacts negatively, too. What is clear is that catastrophes are very costly to a company, no matter how good its previous safety performance. The spectacular has a way of being the only thing the public remembers. Years of building trust may be lost in a short episode. The loss of trust often translates into loss of business.

Cause and Effects

The causes of catastrophes seem simple, although most involve a series of errors. When the release of energy is very large, or there are highly toxic volumes released, and there are people unprotected nearby, the results are predictably severe. The actual results are somewhat deterministic, that is, they depend on time-and-space relationships. The proper combinations for disaster can work both for and against any one individual. For this reason, it is so important to attack the near "hits" just as seriously as the disastrous occurrence. A good system safety analysis can identify the precursors to catastrophe along with the controls to prevent them from taking effect.

A single catastrophic event may sway the public much more than a series of smaller events with comparable losses of life and dollars. One large public transportation crash, either an airplane or train or bus, causes great outrage, yet nearly 50,000 people are killed each year in U.S. traffic crashes, with alcohol being a causal factor in about fifty percent of fatal accidents.

The catastrophic loss of the Space Shuttle Challenger did severe damage to the National Aeronautic and Space Administration's (NASA) image and caused the expenditure of many dollars to convince the public and Congress that it was safe to return to flight. Loss of another shuttle, despite known risks, may threaten major programs, such as the Space Station program.

As Peter M. Sandman has said, outrage is a normal public reaction to things other than hazard in the safety engineering sense. He uses the equation "People's Risk = Hazard + Outrage." The public usually overestimates risk, based upon outrage. Industry thus must do a better job of managing outrage (Sandman, 1990).

Control

Aside from the obvious practice of safety engineering and, in particular, system safety engineering, and the development of protective regulations and standards, an underlying method of catastrophe control would be to better educate engineers. To this extent, the American Institute of Chemical Engineers (AIChE) established a Center for Chemical Process Safety to give expert leadership and focus to engineering practices and research. Safety, health, and loss-prevention curricula materials were developed for chemical engineering schools. Technical and engineering based textbooks with extensive examples of problems and solutions were developed (Kubias, 1989).

CONFINED SPACE SAFETY

Deaths and serious injuries related to confined spaces are a serious and continuing problem. Some 300 fatalities each year are estimated to occur in confined space work (Allison, 1991). The National Safety Council reported that 2.5% of all occupational fatalities were associated with work in confined spaces (Campbell, 1990). OSHA first attempted directly to regulate confined spaces via a 1979 NIOSH Criteria Document call for action. State OSHA confined space standards already existed in California, Michigan, and Virginia when OSHA issued its standard. After a 17-year rulemaking process, 29 CFR 1910.146, "Permit-Required Confined Spaces for General Industry," was finally issued. The standard defines confined space, permit-required confined space, and nonpermit spaces, and unacceptable conditions (Rekus, 1990).

The most common scenario is that one or more workers enter a vessel, lose consciousness due to an undetected contaminant, and are discovered later, often with follow-on multiple fatalities. Statistically, 60% of the victims are would-be rescuers (Colonna, 1987). Some 35% of OSHA-reported fatalities were due to oxygen deficiency. Fully 40% of confined-space victims are fatalities.

Hazards

Confined spaces are difficult to define precisely and thus cause regulatory difficulty. Some common hazard attributes include lack of adequate ventilation, difficulty of access and/or egress, possibility of accumulation of toxic or flammable contaminants or lack of available oxygen, and nonnormal occupancy. Hazardous atmospheres may be classified into four categories: flammable/explosive, toxic, irritant/corrosive, and asphyxiating (Pettit et al., 1987).

Hazardous energy in the form of rotating machinery, electrical sources, extreme temperatures, radiation, and liquid/solid flows are also of concern in confined spaces. Specific energy-control guidance is given in NIOSH publication 83-125 (DHHS [NIOSH], 1983).

Confined spaces should be further classified as to their criticality (frequency and duration of access, severity of hazards, and number of persons at risk). An evaluation of these factors should suggest relative importance for control.

Complete confined-space listings and good identification on maps is necessary for comprehensive confined-space programs. Richard Pearson has suggested a coded numbering system for confined spaces that indicates location, hazard class, and exposure cues in the number (Pearson, 1987).

Some confined spaces present physical and psychological difficulties in accessing the space (deep shafts, tight clearances, tortuous pathways, compounding factors such as poor visibility, slippery surfaces, cold/heat, noise/vibration, dust, odors, wind, remoteness, high elevations, lack of room for multiple persons, and unavailability of assistance). Carefully engineered, innovative methods of access and rescue are often needed. Power winches, body harnesses, or other suspension devices are common confined-space equipment.

Controls

Controls of hazards in confined spaces are similar to that outside confined space. Exacerbating factors include limited air volumes, limited ability to escape from proximity hazards, inability to control environmental conditions such as temperature, and inability to remove toxic materials other than with a confined-space entry.

Common confined-space hazard controls include automation of entry preparations, ventilation, washing or purging of spaces, deenergizing electrical equipment, automated monitoring/alarm systems, redundant critical systems, buddy systems, detailed operating procedures, emergency response planning, training, and equippage, and, finally, warning signs and motivational campaigns. NIOSH, the American National Standards Institute (ANSI), and the American Society for Testing and Materials (ASTM) all advocate keeping combustible gas levels below the 10% of lower flammability limit (Campbell, 1990). It should be noted that a medical surveillance program may be necessary relating to medical demands of confined-space tasks, but it would not be unique to confined spaces.

Documentation

Written confined-space procedures are important as a basis for operating hazard analysis and safety performance auditing. Entry/work permits can be helpful as checklists and aids to documentation. It is important that confined-space tasks are clearly described both as to function and responsibility, for normal and emergency situations. Training involves the necessary education of both the worker and the supervisor. Monitoring provides assurance that safe operating conditions and requirements are met. Industrial hygienists are often involved in the assessment of toxicological/physiological aspects of confined spaces.

Confined-space programs are subject to the old adage, "The job isn't complete until the paperwork is done." Programmatic or legal assessment of performance demands good records. Regulators correctly admonish that "if it wasn't documented, it probably wasn't done," and likely cannot be proven in court. Due to health implications of chronic agents, sampling records associated with confined space work should be retained indefinitely and not be subject to routine documentation purges.

CONSTRUCTION SAFETY

Construction safety involves a wide variety of hazards that are made worse by field conditions and incomplete structures that lack safety features. Large energies may be involved in construction due to large masses and heights. Construction worker turnover may be high and workers may come from large labor pools with varying degrees of knowledge, training, experience, and motivation, making it difficult for them to adhere to safe work practices and rules. Environmental conditions often adversely affect already hazardous work.

Construction safety statistics indicate injury and death rates are much worse than average for industry in general. According to Bureau of Labor Statistics (BLS), only 5% percent of the nation's workforce has 20% of all traumatic occupational injuries (Hislop, 1991). The injury/illness rate of 14.3 per 100 full-time workers (1989), makes construction America's most hazardous industry (La Bar, 1991). The all-industry rate is less than 8.5. Demolition typically has a very high incidence rate. A recent OSHA report, Analysis of Construction Fatalities, indicates that special-trade contractors (electrical, plumbing, heating, ventilation and air conditioning (HVAC), roofing and sheet metal, structural-steel erection, wrecking, demolition, and excavation) accounted for 53% of deaths investigated from 1985–1989, while remaining deaths were in "heavy construction" and "building construction" (*OSHA News*, 1991). The Business Round Table's Construction Industry Cost Effectiveness Project reports that accident costs represent 6.5% of industrial, utility, and commercial construction costs (Hislop, 1991).

Trenching/Excavation

Most construction sites will entail trenches or excavation for foundations, footings, or utility runs. About half of trenching and excavation specialty companies experience cave-ins (Stanevich & Middleton, 1988). OSHA has instituted a "special emphasis" inspection program of all known trenching/excavation projects.

OSHA estimates about 100 fatalities per year due to cave-ins, primarily from trenching. This is about 1% of all U.S. work-related annual deaths. The trenching fatality rate is 112% of the overall construction fatality rate (*OSHA News*, 1988). Death and serious injury rates are particularly high for shallow utility trenches (less than 10 feet deep) (Yokel & Ching, 1983). In a recent five-year study of trenching deaths, 79% occurred in trenches less than 15 feet deep (*OSHA News*, 1988). Furthermore, OSHA has recommended that the present five-foot trench "confined space" definition in its CFR 1926 Subpart P regulations be lowered to *four* feet (OSHA, 1987). It is important to understand that cave-ins occur because the *center* of trench walls fail, not the top edge!

Current OSHA standards require shoring of trenches more than five feet deep and where soil conditions favor collapse. Groundwater in the soil and heavy loads near the side of a hole (such as from nearby equipment operation), and footings require additional protection. Spoil materials shall be kept at least two feet from the edge of trenches. The allowable angle of repose, the slope that the soil will stand without collapse, varies from 2:1 (horizontal to vertical ratio) to 1:1, depending upon the material (loose sand to solid rock). Horizontal benching cuts (steps) are allowed in hard, compact soil, so long as a slope ratio of ¾:1 or less is maintained (26 CFR 1926). OSHA's move toward performance standards now allows four options for sloping and four options for shoring. Timber shoring, hydraulic shoring, trench jacks, air shores and shields per manufacturers' tabulated data, tabular data approved by a registered professional engineer (RPE), and other designs approved by RPEs are now allowed. Excavation work now calls for RPEs rather than "qualified engineers" (*OSHA News*, 1989).

Vertical shoring consists of (1) sheeting to resist lateral pressures of soil and water, (2) horizontal shoring braces to resist compression forces, (3) horizontal stringers to connect or reinforce sheeting, and (4) cleats to distribute bracing forces between sheeting and stringers. The typical shoring arrangement may be simplified by hydraulic shoring devices which expand under hydraulic pressure, capable of high compressive loads, as illustrated in Figure 4-2.

Trench boxes are a prefabricated shoring structure used to provide worker protection and to hold excavations open for the placement of piping and wiring. These devices are often bulky enough to warrant mechanical handling. They also may allow back-filling around a free-standing structure where extremely poor soil conditions do not allow a trench with well-defined sides. Access to the bottom of deep trenches, both in standard shoring and

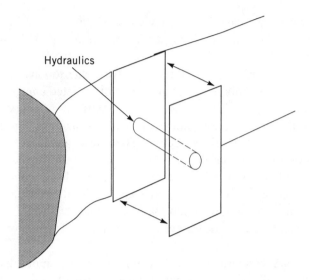

Hydraulics

FIGURE 4-2   Hydraulic Shoring

trench boxes, may be difficult where long sections of pipe or other materials have to be handled, and where slippery conditions are common. Access for both personnel and materials should be designed and provided to avoid falls and such ill-advised practices as bucket riding or crawler backhoes inter-locking booms in order to make it out of trenches. Emergency access for rescue must be provided for any employee in a trench.

Personnel protection equipment (PPE) shall be utilized for recognized construction hazards that are not able to be eliminated. Since overhead hazards usually exist in excavation work, hardhats are a must. Depending upon conditions, gloves, boots, coveralls, eye protection, hearing protection, respiratory protection, and other whole-body protection may be required. Safety equipment such as harnesses, lanyards, and ropes with appropriate fasteners, and rescue baskets may be required for access in normal and emergency situations.

A competent person must design and monitor excavation and shoring. A study of 85 OSHA-reported cave-in fatality cases over a seven-year period revealed that top-width dimensions were only 60–65% of the required value where sloping was involved (Stanevich & Middleton, 1988).

Similarly, a competent person must be involved in respiratory protection programs. Daily excavation worksite inspections shall check for evidence of ground cracking or caving. Spoil or backfill material often obscures the area where cracks develop. Spoil piles above four feet high must be sloped or shored themselves, to prevent them from sliding back into the excavation.

Air quality shall be checked in excavations where toxic materials and oxygen deficiency may be encountered, such as hazardous waste site work.

Electrical Hazards

Seventy-four percent of electrical shock deaths in construction were caused by currents over 480 volts, with 65% involving overhead power line contact (*OSHA News*, 1989). Asphyxiation and fibrillation of the heart can occur based upon voltage and current flow. Moisture, which is commonly present at construction sites, helps current flow, adding to electrical hazard. Many construction tools and field conditions can damage electrical systems. Let-go thresholds are as low as 10 milliamps and fibrillation thresholds are as low as 75 milliamps.

Electrical work other than routine connections/disconnections and switching shall be limited to qualified personnel. Unauthorized modifications to electrical systems shall be made a suspension offense. Voltages need to be measured by qualified personnel. If there is doubt, have the voltages read.

Temporary wiring, use of extension cords, and operation in wet conditions is common at construction sites. Use of grounded or double-insulated tools and ground-fault interrupters is advised.

Power distribution shall be designed and approved by a qualified person. All voltages and their controls shall be appropriately and permanently marked. Voltages 208V and above should have physically different receptacles to avoid inadvertent connections. Regular inspections of plugs and receptacles shall be made to assure differentiation. All voltages above 600V shall be clearly identified. All underground power distribution shall be clearly marked for future work.

Panel covers, switch covers, and receptacle covers shall be installed when a circuit is to be left energized.

Protection shall be taken against contacting the "fishing" end of a metal snake, used in conduit wire-pulling operations, with any adjacent live circuits.

Standardized tagout of defective electrical equipment and circuits shall be required along with tagout for tie-in or other work requiring power to remain off.

Elevated Work

Work at elevation is a high-hazard aspect of much construction. Fall protection is a specialty that increasingly involves high technology.

A programmatic approach to fall protection is necessary in construction. A comprehensive plan for recognition, evaluation, and control of fall risks is needed to achieve organizational expectations.

A standard tie-off of a worker using a safety belt and a 6-foot lanyard protects against an unrestrained fall from elevation but creates increased risk of serious injury from impact forces transmitted through the belt. Full-body harnesses, shock-attenuating devices, retractable lifelines and descent devices, climbing devices, nets, access equipment, and barriers are other fall-protection technologies. These are treated in detail in the fall-protection section of safety skills.

Fire Hazards

Construction sites often involve considerable fire potential due to large stockpiles of flammable building materials, poor rubbish-handling practices, availability of ignition sources, unattended periods between operations, lack of automatic detection, and lack of fire-suppression equipment.

Careful planning and execution of materials storage can reduce fire hazard. Highly flammable materials should be segregated and protected, including fuels, plastics, and paints. Waste handling is a necessary part of all construction and, if done poorly, is often a contributor to fires. Construction ignition sources include roofing pots, welding, faulty electrical tools, plumber's blowtorches, gas heaters, smoking, burning of trash, fires for heat, temporary wiring overloads, drop lights, arson, and vandalism.

All ignition sources should be identified and care should be taken to eliminate unnecessary risk. Hot work permits should be used for all welding and open flame construction tasks. Post-shift fire walkdowns are an excellent method of detecting latent fires, especially where welding is involved. Off-hours security should be specially trained to look for and deal with fires. Considering the value, fire detectors and temporary fire-suppression systems may be justified. Portable fire extinguishers should be properly distributed and workers should be trained adequately in their use. If not already available, fire mains should be extended and hydrants installed for major new construction. Long-range fire monitors can be installed to cover construction until sprinklers can be made operational.

Materials Handling Safety

Materials handling hazards are not unique to construction, however, the activities may not be as standardized as in other industries, due to varying conditions. Some of the mechanical material handling involves very heavy loads, loads that are difficult to rig and keep balanced during the move, and loads that must pass over or in close proximity to other workers or the public.

Construction Lasers

Construction lasers are used routinely for precision alignments. The following safety checklist addresses the major issues:

1. Qualified and trained operators (installation, adjustment, and operation)
2. Eye protection required over 5 milliwatt power levels
3. Area use placards
4. Beam shutter or cap when left unattended for substantial periods
5. Mechanical or electronic alignment vice "eyeballed" alignment
6. Beam not aimed at employees
7. Environmental distance restraints (snow, rain, dust, fog)
8. Label of maximum output
9. No direct staring at beams over 1 microwatt per square centimeter
10. No incidental observation above 1 milliwatt per square centimeter
11. No diffuse reflective light above 2.5 watts per square centimeter

Contracts

In addition to the implementation of physical controls, construction safety requires advanced detailed planning. Hazards, risks, and their controls must be developed before the work begins. Contracts must be worded to require specified control actions. Specific requirements for safety practices, inspection, training, safety reporting, and emergency response are appropriate to contractual arrangements. General wording to obey all safety laws and regulations is not sufficient direction.

CONTRACTOR SAFETY

Contractor safety, not to be confused with construction safety, involves the control of a contractor's (or subcontractor's) safety performance, often in a multiemployer setting. With today's trend toward using more contractors, contractor safety is a programmatic skill that requires both managerial and technical skill of the safety engineer who oversees contractor safety programs for the prime company. The contractor organization provides safety engineering like all prime businesses, so that aspect is not being discussed. A very common problem is that contractors are not managed by the prime with regard to the prime's responsibilities for actions of its contractors and subcontractors.

The level of control often is limited by contractual language. However, it is important to recognize that the contract issuer has responsibilities for its contractor's performance. This is an implicit duty to take reasonably prudent actions to assure that contractor actions do not cause harm, similar to the responsibility it has for actions of a representative of its own company. Although some liability may be delegated away by contract, there remains an underlying duty to select a qualified contractor and to ensure that generally accepted safe work practices are ascribed to and followed. Not to do so may be a negligent act.

There is contention at present as to whether OSHA regulations establish a standard of care due third-party non-direct employers. Gary D. Smith describes the complex legal issues as follows. The employer clearly bears responsibility for compliance with OSHA regulations, but who is "an employer"? The act imposes specific regulations and general duty to provide reasonable care to protect workers from recognized hazards likely to cause death or serious bodily injury. Case law has to establish exactly what standard of care is required of general contractors in a multiemployer situation, and what is required of the subcontractors who are also employers. OSHA was not designed to provide a basis for independent causes of action in civil court. However, violation of OSHA regulations does provide evidence as to the standard of care given employees. In a contract, OSHA regulations form the basis for third-party liability, but establishment of negligence must be decided by courts, often using expert witness testimony. "Negligence per se" generally depends on the "reasonable man" conduct standard which relies exclusively or in part on promulgated regulations (G. Smith, 1991). Resolutions will be settled in differing matters, in differing ways, by different courts. I do not believe that companies should plan to seek refuge behind a chain of employers, because plaintiff lawyers will select those with the greatest ability to pay, and those at the top who austensibly have the most to lose by laissez-faire methods.

The MIL-STD-882 System Safety standard and many informed contracts require specific contractor safety program actions to satisfy this type of duty. Contracts may also specify data requirements such as specific reports, specific formats, required distributions, and so on. It is important that contractor programs be tailored to the scope and sophistication required.

Not only should there be requirements for the contractor tiers below the prime, there should be monitoring, surveillance, or auditing of contractor performance against the safety requirements just as other contract requirements should be tracked. DuPont (1991) has published the ten elements below that are found in better-than-average contractor safety programs. These were developed in conjunction with a 1970 Business Roundtable report that was updated with a 1991 supplement.

1. Know the contractor's safety history.
2. Develop safety goals for contractor safety performance.
3. Evaluate the contractor's safety department.
4. Define safety as an integral part of contractor supervisor responsibilities.
5. Provide trained personnel to administer and follow the contractor's safety activities.
6. Use a system of work permits for potentially hazardous activities (flame, confined space entry, lockout, and so on).
7. Conduct audits of the contractor's safety activities.
8. Collect, maintain, and analyze contractor safety statistics and chart the progress.
9. Conduct investigations of contractor injuries and serious incidents to prevent recurrence. Publish and disseminate investigation findings throughout your organization.
10. Require the contractor to do the following:
    • Establish and promulgate safety policies and procedures
    • Conduct safety training programs for employees
    • Audit compliance to site rules, practices and procedures, as well as government regulations and accepted industry practices.
    • Report OSHA-recordable injuries within 24 hours
    • Investigate injuries and serious accidents

The bottom line for safety engineers regarding contractor safety is that this skill area requires close coordination with legal staff and it is an area where many companies have not received good input from their safety engineers.

CRYOGENIC SAFETY

Cryogenics are liquified gases that must be maintained at extremely low temperatures (<−258°F/−150°C) and/or high pressures to keep them from vaporizing. The most common way to store cryogenics such as krypton, methane, oxygen, argon, carbon monoxide, nitrogen, neon, hydrogen, and helium (in decreasing order of boiling points) is to utilize a Dewar container. Dewars are double-walled containers with a vacuum jacket surrounding much of the liquid, like a Thermos™. Often, a glass surface is silvered to reflect incoming radiant heat. Heavy-duty Dewars are made of corrosion-resistant metals such as stainless steel and have piping and valves which allow gas or liquid to be removed or added. In addition to insulation, Dewars may make use of evaporative cooling to keep the bulk of the cryogenic liquified.

Hazards

Obviously, extremely low temperatures can rapidly freeze and destroy living tissue, depending upon the amount and time of contact. Boiling, splashing, or spilling of cryogenic liquids are to be avoided. Rapid insertion of warmer solids and liquids into cryogenics can cause splattering and boiling. Too-rapid operation of control valves can cause unexpected flows and splashing.

Similarly, these low temperatures can have undesired effects on materials such as metal embrittlement, loss of strength, change of structure, and change of state. Only a few materials are suitable for cryogenic service such as austenitic stainless steels, copper, and certain nickel steels and aluminum alloys. Only a few polymers are suitable for cryogenic service.

One of the often unexpected change-of-state effects of cryogenics on moist air is that of freezing out moisture; and in the case of cryogenics that are colder than that required for air liquifaction ($-318°F$), oxygen-enriched liquid air results.

Ice plugs, frozen out of the air in discharge and relief valves, have caused container explosions. Cryogenics display the characteristically large volume increases (from 1:693 to 1:1,445) that occur during the change from cryogenic liquid to gas. Adiabatic compression of the formed gas can raise the temperature above autoignition points for the three flammable cryogenics: methane, carbon monoxide, and hydrogen.

Fretting of metals in high-pressure, high-flow hydrogen-oxygen cryogenic propulsion systems has caused explosions.

None of the gases resulting from cryogenic liquids are visible or have identifying odors. Only the fog condensed out of the air by cryogenics and/or their extremely cold vapors can be seen.

Controls

Many excellent standards exist and should be used in specifying proper materials for cryogenic service (see the references section).

Personal protective equipment (PPE) should be used to protect against skin and eye contact with cryogenic liquids and their extremely cold vapors. Note that few PPE materials will protect against immersion in cryogenic liquids. Therefore, careful procedural controls are critical. Perhaps the most critical PPE is that for eye protection. Safety goggles and face shields are recommended for splash hazards. An important aspect of PPE is that it should not retain the cryogenic liquid in contact with tissue, and therefore should be easy to remove if it traps liquids.

Special firefighting procedures (beyond the scope of this text) should be followed for fire-related cryogenics.

ELECTRICAL SAFETY

Hazards

Electrical hazards cause many deaths and serious injuries because of wide exposure and broad use of this energy source. Electrocution is the fifth leading cause of death from injury, with approximately 1,000 per year and more than two-thirds of these are work related. Construction and contact with high-voltage power lines, such as with crane lifts, are most often involved. Significantly, the largest group of non-electrical work electrocutions resulting from contact with high-voltage power lines were caused by proximity. Interestingly, about one-quarter of those killed working on high-voltage alternating current (AC) lines were in insulated aerial buckets (Seruda, 1988).

In the past, OSHA electrical violations were cited 500% more than the closest rival (Kleronomos, 1976).

Many of the electrical hazards have to do with poor procedural practice, rather than faulty equipment. Yet, in the case of inadvertent contact with energized overhead power lines, detection of proper clearance distances was unreliable by highly skilled and experienced crane operators and clearance observers, in large part due to lack of adequate visual cues (Cunitz & Middendorf, 1985).

Some 70% of electrical fires investigated in a large study resulted during short-circuit conditions that did not quite trip the circuit breaker (*Hazard Prevention*, Jan./Feb. 1987). Fred Franklin states that North American circuit breaker design (lower sensitivity to magnetic fields) allows too much let-through current once the short circuit arcs. Studies showed that 120-volt arcing circuits normally generate 200–250 amperes of flow (150–400 Ohm range) that are fully capable of melting/welding the conductors. Furthermore, the arcs extinguish and reignite repeatedly, causing missing half and quarter cycles in the sine wave. Adoption of solid-state sensors/computer activation or change to the European 9-coil circuit breaker design could reduce fires by 20 percent (Franklin, 1990).

Where it was allowed, aluminum wiring in household branch circuits has caused fires due to the increased resistance associated with oxidation, despite circuit breaker protection. Most fires due to flying metal globules were contained due to the use of junction boxes for connections.

Controls

Intrinsically safe electrical equipment, such as double- and triple-insulated hand tools, and low-voltage systems are good examples of engineering controls.

Grounding of portable electrical equipment is a problem because a short to case can generate 12 amps of flow with a cumulative grounding resistance of 10 Ohms. The typical branch circuit breaker will not trip until 15 amps, potentially electrocuting the user. Furthermore, circuit breakers are designed only to protect equipment, and are allowed a maximum tripping time of 2 minutes at 200% of the 0–30 amp rated current. The let-go threshold is only 10–12 milliamps and fibrillation can occur between 60–125 milliamps (Kleronomos, 1976).

Ground Fault Current Interrupters (GFCIs) are designed to open the circuit before a fault path through the operator can cause harm, as low as 5 milliamps. Complaints are raised about the impracticality of GFCIs due to the many false trips, especially where long wire runs are involved and damp conditions exist. The two kinds of GFCIs are the differential transformer type and the isolating transformer type. The latter can limit flow to as low as 0.2 milliamps (below the perception threshold) (*National Safety News,* July 1980). GFCIs or an Assured Equipment Grounding Conductor assurance program are required for receptacle outlets in construction. Ground and continuity monitors have been designed to provide a visual indication of proper grounding continuity, which could aid the assurance program.

Grounding and bonding are two hazard controls that, if done properly, work without the need for personnel action at the time of interaction. Regular verification of proper grounding and bonding is necessary to assure presence of the controls. Such testing can be done with proper resistance measurements. Improper indications of continuity of ground or neutral can occur using neon testers, where neutral or ground wires couple (capacitively) in long wire runs (Kleronomos, 1976). A commonly observed safety discrepancy is the deliberate removal of the grounding prong from plugs, so that ungrounded extension cords or outlets can be used. Also, adapter plugs are improperly used, without connecting the pigtail to a proper ground.

Lower in the hazard-control precedence order is the use of personal protective equipment, such as insulating gloves and tap sticks. Even lower is the sole reliance upon training and procedural protections. This is not to suggest that such hazard controls are not valid; they are less desirable controls and require a high amount of effort to accomplish with certainty.

Electromagnetic Fields

Recently concern has been raised about electromagnetic fields (EMFs), which are associated with 60 cycles per second AC flow. Conflicting information has appeared supporting both no-hazard and suspected health hazard positions. The Electrical Power Research Institute (EPRI) has 30 studies under way in this area (La Bar, 1990). The federal Environmental Protection Agency (EPA) has a draft report that reflects an Office of Technology Assessment statement that emerging scientific evidence "no longer allows one to assert that there are no risks [of cancer due to electromagnetic radiation]." But it does not provide a basis for asserting that there is a significant risk (*Occupational Hazards*, July 1990). Barrett Miller discussed the EMF effects issue regarding 100 milligaus to 12,500 milligaus EMF levels in household situations, and several studies, starting with the 1981 Denver study that suggested a 3–5 milligaus cause for 1.7–2.1-times increased cancer risk in children. Study validities and replication have been questionable. Other possible EMF effects include brain tumors, depression, learning disabilities, slower reaction times, and reproductive problems (Miller, 1989; La Bar, 1990). A major problem with the previously suggested associations between cancer and Extremely Low Frequency (ELF) radiation is the lack of definitive exposure information (Anna, 1989).

Other Electrically Related Safety Effects

Other undesired safety effects can occur from power supply idiosyncrasies including blackouts, brownouts, sags, spikes, transients, frequency variation, surge, overvoltage, and electrical noise (electromagnetic interference or EMI and radio frequency interference or RFI). These effects can damage both electrical and electronic devices. These problems boil down to not enough power, too much power, and other power aberrations. Edgar Coudal gives these eleven solutions to power line problems (Coudal, 1988):

1. Surge suppressors
2. Shielded wiring
3. Dedicated lines

4. Isolation transformers
5. Voltage regulators
6. Line conditioners or filters
7. Standby generators
8. Motor generators
9. Standby power systems (SPS)
10. Traditional uninterruptible power systems (UPS)
11. Ferroresonant UPS.

EMERGENCY RESPONSE

For the purposes of this book, Emergency Response (ER) is defined as a limited response to abnormal conditions expected to result in unacceptable risk requiring rapid corrective action to prevent harm to personnel, property, or system function. Although ER is a reaction, and engineering tends toward prevention, ER is a skill area that safety engineers must be familiar with both because of regulations and good engineering practice. ER should be designed to cover those situations where normal controls have failed to provide safe conditions.

Also, for our discussions, recovery or restoration applies to the restoration of normal conditions after the emergency is under control; it is not discussed here because it entails moving from a controlled emergency state all the way back to a normal condition. For example, if a fire occurred, ER might entail getting personnel to safe locations, controlling the fire, and limiting the damage—otherwise moving to an acceptable system state. Recovery, then, may proceed to treat long-term injuries, obtain replacement manpower, correct damage, and take operational steps to regain full, normal function—not a usual function of the safety engineering. It is recognized that some short-term functions, such as first aid and rescue, are ER actions, but safety engineering may not be expected to be the provider of such services.

Since "emergency" is difficult to define, the following attributes are helpful: a usually distressing condition that often can be anticipated or prepared for but is seldom exactly foreseen, and a state that calls for immediate action. Three prime examples of emergencies can be given—fire/explosion, natural disasters, and medical crises. Shades of gray in recognizing emergencies can only be settled by local rules.

Phases of an emergency are:

1. Recognition of an emergency condition
2. Assessment and signaling of emergency condition

3. Emergency response
4. Termination of emergency condition

Developing an ER Plan

ER planning is required by several acts, including the Occupational Safety and Health Act (OSHA), the Resource Conservation and Recovery Act (RCRA), and the Superfund Amendments and Reauthorization Act (SARA).

ER as an integrated function may involve many tasks and many organizations, as shown in Figure 4-3.

Some of the biggest problems associated with ER have to do with command and control issues, that is, who is in charge when, who is to do what, and who is doing what. Turf battles such as those between firefighters, emergency medical services (EMS), and police, are classic. To minimize this and other problems an adequate ER plan is necessary.

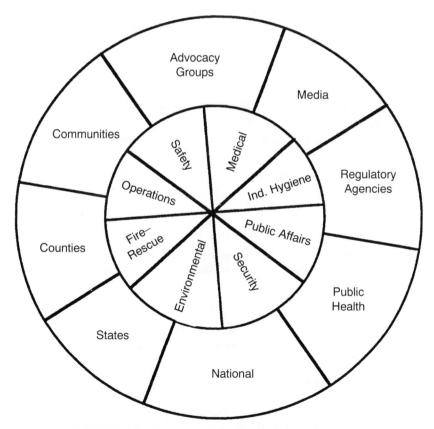

FIGURE 4-3   Emergency Response (ER) Relationships

Minimum elements of an ER plan include:

- General description
- Alert procedures
- Emergency action organization
- Emergency procedure(s) details
- Emergency training requirements
- Emergency maps and diagrams
- Emergency plan updating requirements

The heart of the ER matter can be summarized as

- Protect
- Communicate
- Control
- Record
- Follow-up

Protect

Immediate notifications: key functions to be notified immediately upon recognition of emergency (fire, police, management, and so on)
Minimize risks: protective actions for workers, public, facilities, property
Coordination of response: integrating inside and outside response

Communicate.   Give needs and status to news, staffs, administrators, enablers, suppliers, etc.

Control

Restore: achieve near normal conditions, minimal operations
Cleanup: housekeeping with consideration for fact finding

Record

Fact find: gather transient facts, consult with eyewitnesses
Report: investigative report and recommendations

Follow-up.   Implement corrective actions

Prediction ability is part of the emergency response limitation. Dollars and politics are other considerations not discussed here. Informed generation of scenarios has been expanded to include modeling techniques that are often aided by use of computers.

Gas dispersion modeling has been developed to predict lighter-than-air and buoyant hot gas releases such as in the nuclear industry, and heavier-than-air dense gas releases such as in chemical industries. Even after some time and upgrading, the validity of these models is questionable due to the necessary assumptions and ground rules within the models. Goyal and Al-Jurashi (1990) list more than 100 models. They discuss the instantaneous versus continuous sources, gas jet and liquid release modes, and two-phase flow involved in superheated liquid releases. Further, wind, solar radiation gains and losses, and other heat conduction gains and losses significantly affect the calculation of effects. The authors suggest that multiphase jet sources, better 3-D modeling, better time-variant release models, dispersions with varying confinement, continuous low releases of heavy gases, and dispersions beyond barriers represent some needed improvements.

First Aid in Emergency Planning

Safety departments are often tasked with first aid functions in emergency response and routine situations where medical personnel are not available. Proper training, qualification, and certification in key skills is necessary. CPR (cardiopulmonary resuscitation) and EMS (emergency medical service) skills require special training and certification. Safety programs must place proper emphasis on lifesaving actions. Staffs need to assure adequate personnel for all operational periods. There should be coordination with medical professionals for first aid program guidance and oversight. Improper actions by unqualified personnel can cause increased liabilities to both individuals and companies.

Detailed first aid programs are beyond the scope of this text. Safety programs should review programs such as those offered by the American Red Cross, to evaluate the application of a detailed first aid program, according to workplace needs.

A first aid kit might include these items:

First aid manual
Sterile gauze pads
Sterile gauze tape (1 in.)
Triple antibiotic cream
Adhesive paper
Bandage scissors
Thermometer

Tweezers

Inflatable splint kit (arm, leg)

Cold packs or ice bag and a ready ice source

Cloth squares for slings, and other uses

Elastic-adhesive bandage strips (including knuckle strips)

Sterile sewing needle (for splinter removal)

Betadine solution

Aspirin and aspirin substitutes

The Red Cross suggests this four-step action plan in every emergency situation:

1. Survey the scene.
2. Do a primary survey of the victim.
3. Phone EMS services for help.
4. Do a secondary survey of the victim.

Survey the scene.   Existing hazards may threaten the rescuer as well as the victim. Take proper protections, especially regarding toxic atmospheres. Moving the victim can be more dangerous than existing hazards. Look for potential causes of injury; check for the number of victims who may need help; ask bystanders what happened; identify yourself as a trained rescuer; and ask conscious victims for their permission to help them (it is implied for unconscious victims).

Primary survey.   Check the victim's two most important vital signs: Respiration (breathing) and circulation (heartbeat and absence of bleeding). Use the "ABC" mnemonic: Airway open? Breathing? Circulation?

Phone EMS for help.   Ask two or more bystanders (backup) to call for help. Call 911 or operator (look on pay phone label or inside directory cover). Tell where (phone number/location), who (caller), what (what happened) and how many injured/care being given.

Secondary survey.   Interview victim. Check for normal breathing, pulse, body temperature. Check body head to toe for additional injuries.

First Aid Care

First aid provides initial care, especially life-sustaining care, until EMS personnel or advanced care professionals can assist the injured. The following areas need first aid preparation to provide care:

1. Heart attack, stroke, fibrillation, electrocution
2. Lack of breathing
3. Unconsciousness
4. Profuse bleeding
5. Serious burns
6. Shock
7. Serious eye injury
8. Heat stroke
9. Seizure
10. Drowning

Lack of breathing.   Lack of breathing is life-threatening because brain damage may start if oxygen is interrupted over eight minutes. Some chemical exposures further complicate the return of breathing by blocking the uptake of oxygen in hemaglobin.

The old back pressure, arm lift method has been shown to be much less effective than mouth-to-mouth breathing. The fear of communicable disease, and contact with objectionable fluids, has given rise to a variety of breathing aids to avoid direct contact with the victim. Effectiveness varies. If air exchange is not accomplished, the devices should be abandoned in order to save a life. Positive-pressure oxygen-delivery devices can improve the ability to revive victims, such as with anhydrous ammonia exposures where oxygen exchange is impaired.

Where oxygen uptake is blocked by such chemicals as carbon monoxide or cyanides, administration of intravenous drugs (qualified medical professionals only), may be necessary in addition to 100% oxygen administration.

Rescue breathing (victim needs help) (American Red Cross, 1987).

1. Check for response (tap and shout "Are you OK?")
2. If not done, shout for help (alert bystanders)
3. Position victim* (straighten victims legs, pull onto back toward rescuer, stretch nearest arm over victim's head, support head, all within 10 seconds)
4. Open airway* (head tilted back and chin lifted)
5. Check for breathing (watch chest rise/fall and feel airflow out of nose/mouth)
6. Pinch victim's nose closed, give four full breaths, watching for chest

---

*Depending on injuries

movement. If necessary clear airway with finger sweep, give abdominal thrusts (no back slaps), look for stomach rising.

7. Check for arterial pulse alongside Adam's apple, using fingers not thumb. Check for 5–10 seconds.
8. Phone EMS if not done (report consciousness, breathing, pulse condition of victim).
9. Begin rescue breathing every five seconds. Every fifth breath, check for return of breathing.
10. Recheck pulse after 12 breaths (once every minute). If no pulse, do chest compressions per CPR training. (Compressions if heart is beating on own can cause injury.) Hand placement, amount of compression, and timing of compressions are critical to CPR effectiveness. Adults need 15 compressions, followed by 2 breaths. Two rescuers are better than one for full CPR.
11. Keep it up until advanced care arrives.

Spinal injury.   About 15–20% of head injury victims have spinal injuries (*Safety & Health*, April 1988). First aid checks for spinal injury include:

1. Ask conscious victim about pulsating pain in ribs, arms, or legs. (If yes, suspect injury.)
2. Can victim move feet, legs, fingers? (If no, suspect injury.)
3. Can victim exert firm pressure with feet and fingers? (If no, suspect injury.)

Try not to move spinal-injury victims.

Profuse bleeding.   Most profuse bleeding can be stopped with direct pressure using a clean cloth and, if necessary, just the hand. If pain is not increased, elevate the wound. Only if bleeding persists, or if arterial spurts of blood are occurring, apply pressure to pressure points per first aid training. Tourniquets are a last-ditch effort where other methods will not work, such as a severed limb. Tourniquets should not be removed by first aid rescuers, and a "T" should be marked on the victim's forehead.

Serious burns.   Serious burns need immediate treatment. Burn treatment has changed in recent times. Cool, clean water will obviate some minor burns if the wound is quickly immersed. Burns that quickly blister should be covered with a clean, wet cloth such as a sheet. Burns that appear white or which have charred the skin should receive immediate specialized burn care.

Heart attack.   Of the million and a half persons who have heart attacks in the United States each year, approximately one-third die, and most of those die before reaching a hospital. Overall, 50% of Americans die from heart attack or blood vessel-related disease (cardiovascular disease). Some heart attacks must be expected on the job.

Heart attacks should be suspected with pain or pressure sensations in the chest. (Sweating, nausea, and shortness of breath are other heart attack symptoms that can result from other causes.) Quick reaction by first aid responders, despite uncertainty or patient denial, improves survivability dramatically, because most heart attack victims die within the first two hours after the onset of the heart attack signals.

If a heart attack ends in cardiac arrest, or other trauma causes heart stoppage, CPR care must be initiated within four minutes, and then be followed up by advanced medical care, preferably within eight to ten minutes. Survival can occur outside these times but, in general, quick and appropriate action optimizes survival.

Stroke.   Factors which may identify stroke include

- Unconsciousness
- Paralysis/weakness on one side
- Breathing/swallowing difficulty
- Loss of bowel control
- Slurred speech
- Uneven pupil dilation

Stroke first aid includes administration of oxygen, treatment for shock (give no fluids), and transportation to advanced care with the victim on their paralyzed side, not the back.

Unconsciousness.   So long as vital signs are normal, unconsciousness in itself is not life threatening. The victim should be treated for shock and advanced care should be obtained as for other serious injuries. If victims recover consciousness, they should see a medical professional immediately because of the chance of concussion or other serious cause of the unconsciousness.

Non-life-threatening injuries.   Almost a third of lost workday cases involve injuries to lower extremities. The more common of these non-life-threatening injuries are sprains to the ankle/knee, toe fractures, and bruises to the lower legs and feet.

A very common first aid need is to treat strains and sprains. Slightly more than half of emergency physicians reported using both cold and heat for strain treatment. Some 97% used cold in initial treatment for the first 48–72 hours, according to the common first aid equation RICE (Rest, Ice, Compression, and Elevation). Cold along with compression seem to give better recoveries. Self-cooling chemical packs are available, as are liquid-to-gas expansion units such as the Cryopac™. Cold applications should be limited to 20 minutes. First aid responders should watch for diabetics, the elderly, or those with circulation impairments when using cold therapy. Controlled heat is used, in the later stages of treatment, after cold treatments have reduced swelling. Warm whirlpools and ultrasound are commonly used in advanced care of sprain/strains. Serious strains and sprains should be referred to medical professionals for evaluation.

Safety engineers, especially in smaller organizations, will be expected to supply first aid expertise and act as first responders in many emergency situations. Preceeding text examples should be refined and tailored to individual industries, using medical professionals.

ERGONOMICS

There are two terms used in the United States for the process of designing for human operator safety and efficiency: ergonomics and human factors engineering (HFE or human factors, for short). Since the British-originated term ergonomics (also known as biomechanics) seems to be more broadly used than human factors, it will be used in this book to refer to both disciplines.

Ergonomics is a relatively new term in common usage, (accepted by Merriam Webster's Dictionary in 1961), drawn from the words for the study of work, an ancient concept with a new name. HFE clearly involves the design process for human use (McCormick, 1970). The Human Factors Society was formed in 1957, mainly centered about industrial engineers and engineering psychologists in the California aircraft industry (R. Smith, 1991). Despite nomenclature differences, the focus of both human factors and ergonomics is generally that of systems optimization of the man–machine interface. Since ergonomics is a systems concept, it encompasses a broad range of subjects and draws on many interdisciplinary skills. Ergonomics addresses both the design and operational aspects of work.

Ergonomics involves such things as anthropometry (physical dimensions and weights of people, physical reaches); human performance in terms of physiology and mental functions; effects of stress upon performance; human error rates; human interaction with machines, especially the computer; tool design; workplace design; task design; control layouts; procedure enhancement; behavioral stereotypes; learning curve applications; handicapped and temporarily handicapped worker accommodation.

Najmedin Meshkati (1990) astutely notes that a good majority of human operator error is attributable to design-induced or forced errors. Human error is an attribute of the whole technological system. Further, operator error is neither necessary nor sufficient to cause many concatenated failures. These are strong arguments against the oversimplification that operator error causes the majority of accidents.

Analyses in the ergonomics arena take several forms, both qualitative and quantitative. One commonly used technique is A.D. Swain's Technique for Human Error Rate Prediction (THERP). This has been expanded by Bell and Swain into Human Reliability Analysis (HRA) by combining a task analysis with THERP to generate Performance Shaping Factors (PSFs). HRAs are used to support the larger Probabilistic Risk Assessments (PRAs) done for nuclear power industries and other complex, high-energy industries. Human performance of concern is often identified by qualitative fault tree analysis. Databases of human error rates for specific tasks are increasing. Earlier databases had a mix of data which often did not closely resemble the tasks to be evaluated. Task error probabilities can be applied, with caution, in fault trees or event trees (discussed further in the system safety chapter). It should be recognized that there still is not a way to use current human error information to predict human reliability accurately.

Jens Rasmussen (1988) notes that automation removes boring, repetitive tasks, placing the human into a supervisory role; it also means that human error is not easily predictable based on elements decomposed from standard routines. Computerized modeling and expert systems should help shed light on cognitive error prediction and task design for reduction of human error. Elimination of all human error is an unrealistic goal, as is zero defect rate. Costs of total error elimination increase asymptotically as the goal is approached. This ergonomically based source of error drives the need to produce systems that are tolerant of some reasonable amount of error (failure/fault tolerance).

Ernest McCormick describes three system types involving man: a manual system (man as power source and controller), a mechanical system (man as a controller), and an automatic system (man as a monitor). The functions in a man-machine system are in Figure 4-4.

Control System

A great deal of ergonomic engineering work is related to control systems. Display instrumentation, such as predictor displays, coupled with computer processing and simulation, has extended the human operator's ability to

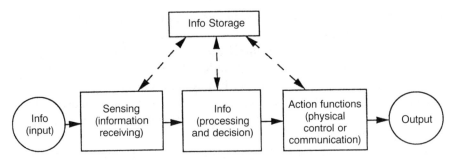

FIGURE 4-4   Man–Machine System Functions. *Source:* McCormick, Ernest (1970) Human Factors Engineering, Third Edition, New York: McGraw-Hill.

predict, diagnose, and control complex systems (Banks & Cerven, 1985). Much of the early command display work comes from ballistics tracking. Early methods relied on computing and display of a recommended tracking response. Spacecraft control has been achieved by incremental on-line control actions, bracketing the desired performance. Off-line control reserves the action, as long as time allows, while the proposed control is evaluated by things like prediction and tracking. Finally, supervisory control takes over control from the human operator, with only a manual override option.

A tremendous amount of study has been given to human perception and the related ability to recognize hazardous conditions through displays. The spatial layouts of modern cockpits and control rooms reflect technology that improves human ability to perceive critical information quickly and properly. Annunciation systems, strip gauges, combined sensory input devices and arrangements of panels are examples of improved information processing based on ergonomic principles. Size and shape of character fonts have been evaluated for optimal recognition. Auditory signals and alarms have been extensively studied. With microprocessors, voice synthesis annunciation is a reality. Speech communication studies in the context of various noise backgrounds are available.

Computerized information processing allows data reduction and trend recognition without overloading operator senses. Much has been written about the thought processes that must precede the taking of control action. Interactive skills and processes such as information storage, memory and retrieval, learning, and the effects of training such as simulation should be researched in designing a man–machine system.

The actual design of controls that the person in the system must recognize and operate has been tested as to shape, location, force requirements for proper operation, and feedback. The biomechanics of work is an area of bioengineering that provides design parameters relating to human capabilities (man as the power source). Extensive NASA studies were necessary to design tasks for astronauts, especially the Extra-Vehicular Activity (EVA) tasks in low gravity and in space suits. Ergonomic studies such as those for manual lifting tasks are discussed in the Back Injury section of this book.

Applications

Ergonomic design.   An important consideration of ergonomic design involves human requirements related to body dimensions. Anthropometry databases, for both static and dynamic ranges, give key dimensional information about the wide variety of human shapes and forms, both male and female. An example of a comprehensive anthropometry database is NASA's series of Man-Systems Integration Standards (MSIS). Design to the 95th percentile of the planned use population is commonly used to accommodate two standard deviations of a normally distributed group, on either side of the mean. As the workforce changes, including more females and older workers, anthropometry values will change, requiring a redesign of work clothing and equipment (see the Personal Protective Equipment section).

An ergonomic mismatch occurs with the 10–14 % of Americans who are left-handed and forced to use right-handed designs. A 1989 British Columbia study of almost 1,900 students found that about 90% of left-handers were more likely to suffer serious accidents. Left-handers are 135% more likely to have car driving accidents, supposedly due to a normal defensive action that raises the right hand high for protection, sending the car into oncoming traffic. Band saws, table saws, and chain saws were also identified as more difficult for left-handed operation. Neutral tools, operable for both right- and left-handed persons seem to be a workable answer (Merriman, 1990).

Ergonomic design also concerns itself with the physical arrangements and characteristics of the work environment, such as heat, noise, vibration, illumination, and other physiological stressors. There is an overlap and interaction between ergonomics (human factors) engineers and industrial hygienists in dealing with workplace stressors.

Engineering disciplines such as operations research can be applied to spatial and task optimization. Video Display Terminals (VDTs), alternatively called Video Display Units (VDUs), provide interfaces between humans and computers. Ergonomic applications to VDTs are a good example to illustrate the system functions already discussed.

VDT/VDU effects.   A scientific assessment of electromagnetic fields and X-ray emissions revealed no evidence that VDT operators are exposed to electric, magnetic with Very Low Frequency (VLF) magnetic fields being marginal, or ionizing radiation fields significantly above ambient levels. Also, significant X-ray leakage cannot occur under any credible conditions. Comparing discomforts between VDT operators and non-VDT typists, only eyestrain, blurred vision, and neck/shoulder aches occurred with significantly higher interest. Skin rashes, headaches, backaches, and arm/finger aches were not different between the two groups (Walsh et al., 1991).

VDT ergonomic guidelines give optimal eye-to-screen distance ranges, angles of depression, keyboard position and layout, chair heights, and footrest recommendations. Other ergonomically based guidelines address monitor design (phosphors, character layout, refresh rate), background lighting, glare, time of tasks regarding breaks, fatigue-relieving exercises, and so on.

Cumulative trauma disorders.   Perhaps the most significant new application of ergonomics is to cumulative trauma disorders. Cumulative trauma disorders (CTDs) are often ergonomically related work effects (although there are significant nonwork causes also). They are also described as a musculoskeletal injury and also as repetitive motion injury. For the purposes of this text, CTDs are used rather generically.

It is important to recognize that CTS is different from related cumulative trauma disorders such as tennosynovitis (nerve sheath inflammation) (Hoyt, 1984) and Raynaud's Phenomenon or Vibration White Finger (circulatory disorder) (Goel & Rim, 1987). The normal cumulative trauma disease progression is from muscular tendonitis to nerve disorders, but CTS normally occurs only in a small percentage of those continuing to be exposed to causal factors, compared to those having muscular disorders.

According to the Bureau of Labor Statistics (BLS), CTDs are considered a subset of the much larger "soft-tissue" injury set. CTDs are the fastest growing category of injury today, accounting for more than 30% of all workers' compensation claims (*Occupational Health & Safety*, July 1990). Carpal Tunnel Syndrome (CTS) has received a great deal of emphasis in terms of treatment and compensation for injuries. Interestingly, CTS was identified as early as 1865. And, according to 1979 BLS data, CTS claims occurred in about 20 per 100,000 manufacturing workers, representing about 70% of nonimpact wrist disorders (Hoyt, 1984).

Early CTS symptoms include abnormal sensations such as prickling or itching of palmar surfaces of all but the little finger, and swelling, while later symptoms include pain, numbness, tingling, loss of muscle grip control, and lack of sweating in parts of the hands (Johnson, 1985). Importantly, symptoms increase at night, tend to become more persistent and intense if untreated (Groneman, 1985), and may spread as far as the base of the neck

(Folkers, 1986). Women appear to be affected two to ten times more often than men (Johnson, 1985).

Causes

Repeated, forceful, up-and-down deviations of the wrist or repetitive motions of fingers while holding the wrists unsupported or resting on a corner are generally mentioned as causal factors for CTS. The underlying cause is injury to a nerve as it passes through the carpal tunnel in the base of the wrist. Additionally, direct pressure and vibration may be factors. Nonoccupational repetitive motion causes such as hedge clipping, grass shearing, and vacuuming have been implicated, especially affecting the dominant hand, and non-work disease etiology (Groneman, 1985).

Controls

CTS hazard controls are instructive of the use of ergonomic engineering solutions, personal protective equipment, administrative solutions, as well as medical treatment. Preventive, before the fact, solutions are much preferred over reactive measures which occur after the disorder may have become irreversible.

K. H. E. Kroemer wrote an excellent, comprehensive article for the *American Industrial Hygiene Association Journal* (AIHAJ) describing common repetitive strain injury control recommendations (Kroemer, 1992).

Preventative controls include stretching, flexibility, and strength-building exercises, redesign of tooling (Sawyer, 1987) and, more importantly, ergonomic task/tool redesign (including task diversification, rest periods, automation, tool suspension, tool-holding postural practices, padding workstation armrests, designing for minimum wrist flexion in right angle tools, and vibration damping. Additional ergonomic tool design includes proper size and shape of handles, tool surface texturing, padding of striking tools for shock absorption, and isolation of low-frequency vibrations from 10–60 Hz (Johnson, 1985).

Medical treatment is a post-injury control and ranges from rest, ice/heat treatment, and immobilization or support of the wrist, to anti-inflammatory drugs, to surgery, and vitamin therapy, especially B_6 (Folkers, 1986). Surgery may not work as much as 20% of the time, or be permanent (Folkers, 1986). Normal scar tissue formation may cause renewed pressure on the carpal tunnel nerves. Many occupational physicians recommend conservative, non-surgical treatment for most cases. However, a neurosurgeon who has done more than 800 carpal tunnel release operations (local anesthesia and special surgical tool outpatient surgery) reports more than 90% "good to excellent" recoveries in a five-year study of 200 postoperative patients (Pagnelli, 1990).

Manual Materials Handling Applications

Methods studies.   The application of ergonomics to manual materials handling addresses the largest significant risk of injury and resulting costs in industry. The major effect, back injury, has been discussed earlier. Good industrial engineering methods studies may eliminate up to fifty percent of lifting tasks, by improved process flows and use of handling aids.

Disabled or impaired workers.   Disabled workers are those with a physical or mental impairment that substantially limits certain job performance. In the past the term "handicapped" was applied to define and protect such workers (Krikorian, 1978). Efforts to hire impaired persons and efforts to return previously fit employees to the job after impairment by accidental injury or by illness, show that matching jobs to disabilities requires a great deal of ergonomic information. Detailed information is needed about both the job and about the individual. Rather than wait for the question of disability or impairment, it is good business to determine beforehand the ergonomic limitations of each job. Hiring qualified disabled workers benefits companies in terms of equal opportunity and dedicated, effective workers who can outperform able-bodied workers if the tasks are compatible with their limitations.

Preexisting physical impairment affects second injury compensation, where specifically legislated (Rodriguez, 1980). This provides additional impetus to ergonomically optimize jobs for those who have a preexisting impairment, whether that impairment fall under legislative protection or not.

The Americans With Disabilities Act (ADA), is the most recent (as of this writing) legislation to mandate accommodations for persons with disabilities. Compliance with such acts should be easier with a good safety engineering program that addresses disability issues.

FACILITY SAFETY

Facility safety implies that the safety of the entire facility is being addressed. For this book, "facility safety" means the safety of the facility that supports some operation or operations. Facility safety generally includes the building, utilities, and ancillary systems that are needed to provide a basis for the operation. Since the distinction between facility and the operation is often blurred, an example is helpful. The facility supporting a series of wind tunnels includes the buildings that house portions of the wind tunnels, offices and control rooms, computer rooms, electrical utilities, compressed air systems, HVAC systems, hydraulic utilities, warehousing, general materials handling equipment, and the like. The operation would include the wind tunnel structure itself, motors, fans, computers, shop equipment, electronics, sting supports, models, and other specific equipment unique to the wind tunnels. The exact demarcation between facility and operation includes a variety of interfaces that should be carefully defined.

Facility safety generally does not concern itself with operational procedures, except as they affect the facility at the interface to personnel. If the operation requires certain functional provisions by the facility, their safe provision must be considered. Similarly, if facility functions can cause adverse safety effects upon the operation, this should be identified and controlled by the facility and perhaps cause some accommodation by the operation, depending upon the severity of the effect.

The safety engineering of facilities involves the standard Recognize, Evaluate, and Control functions that characterize all safety engineering. For recognition, many good checklists have been developed to help assess hazards within facilities.

The following is a checklist of facility hazard concerns common to most facilities:

- Offices
- Operations and staging areas
- Power, fuel systems
- Waste systems
- Water systems
- Warehousing/dock systems/storage areas
- Fire protection
- Security
- Maintenance shops
- Machine shops
- Personal services, health/sanitation
- Food services
- Transportation systems
- Supply systems
- Communication systems
- Ventilation systems

Many insurance companies have such checklists. Checklists may be centered around topical subjects such as life safety, fire safety, health concerns, transportation safety, environmentally based safety concerns, and the like. Some checklists have been expanded to use weighting approaches that in part accomplish evaluation. Controls can be a result of analysis, or in some cases standards and requirements can be invoked to accomplish control. But to be systematic, a system safety approach would be best. System safety is treated later in this section.

An example of a checklist tailored to facilities is adapted from a Leo Greenberg (1979) checklist.

- Plant location (flooding, landslides, earthquakes, avalanches, fire threat, safe access by vehicles)
- Plant hazard to community (distance, meteorological conditions, population density)
- Plant structure (wind resistance, seismic load capability, fire resistance, blast resistance, floor loading, computer floors)
- Plant layout (egress and life safety, security, fire zones, isolation capability, segregation of pedestrians from vehicular traffic)
- Plant utilities (siting of high pressure/fired boiler utilities, high-voltage electrical vaults and high-voltage switchgear, large water storage tanks, water cooling towers, compressed air spheres, toxic gas piping systems, overhead high-voltage suspension lines)
- Plant storage (chemical storage, high-stacked storage, flammable materials storage, fuel storage, cryogenic storage tanks)
- Plant transportation systems (elevators, manlifts, scooters, trucks, air pallets)
- Plant material handling systems (mobile cranes, scissor lifts, cable platforms, conveyors)
- Plant emergency systems (alarm/emergency communication systems, fire protection/suppression systems, emergency ventilation, fire doors, smoke vents, explosion suppression systems, emergency evacuation systems, shutdown systems, isolation systems, rescue systems, decontamination systems)

An unavoidable source of facility hazard is deterioration of structural materials because of aging. Often older facilities have unsafe pressure systems (corrosion, abrasion, vibration), improper selection of materials for expected environments, and problems caused by normal wear and tear. Preventive maintenance inspections and actions are vital to avoiding undetected, unsafe physical facility conditions. A ubiquitous source of hazards in facilities relates to the recurring need to upgrade and modify facilities (not all of the upgrades are due to deterioration.) Change, if not properly managed, is a major source of safety problems. Formalized change review by safety engineers is advisable. Construction modifications are a fairly common source of changes.

FALL PROTECTION

Fall protection is a safety concern that refers to controlling falls from elevation and slips/falls from the same level. The NSC predicted 1,200–1,600 deaths and more than 400,000 disabling injuries each year from falls. Falls were the third leading cause of disabling injuries in the United States (about 17% of all workers compensation claims). Based on state data, falls represented from 7% to 13% of all reported accidents. The relative proportion of falls from elevation to falls from the same level ranged from 1.35:1 to 1.70:1 (more falls from elevation were reported, probably more serious effects) (Pater, 1985).

Falls from Elevation

A 1984 BLS 24-state survey reported 60% of elevated falls were under 10 feet and 50% of those were under 5 feet (Pater, 1985).

Another BLS survey reported in the 1985 Accident Facts publication stated that 70% of reported falls were from scaffolds, 14% from roofs, and another 14% from boxes, barrels, equipment, or furniture. Further, 85% of falls from elevation resulted in lost work days and 20% resulted in death (Eisma, 1990).

Obviously, there is a close connection between the walking and working surfaces and the tasks that contribute to falls from elevations. Errors, material and mechanical failures can cause loss of balance and falls. Winds, visual effects of great heights, slippery surfaces, irregular platforms, narrow walkways such as beams, rapidly moving suspended loads, and other distractions are examples of fall hazards that occur at elevation. Lack of barriers, handrails, and restraint in working surfaces and machinery seats can also result in worker freefall. Manlifts, conveyor belts with personnel steps, and hand grips whose continuous motion require workers to time entry and exit properly have a bad reputation for injury and even deaths. The act of carrying items, effects of weather, and temporary lapse of consciousness or balance can all contribute to manlift mishaps. Falls from the same elevation, slips and trips are treated in the Walking and Working Surfaces section where ladders, scaffolds, and work platforms will be discussed.

There is debate about what should define an elevation. For our purposes, we shall use the nominal 6 feet suggested by J. Nigel Ellis, as a compromise between 10 feet and 4 feet (Ellis & Lewis, 1986). An engineering approach to selecting the most appropriate elevation to mandate fall protection might be to identify a significant break in severity plotted against fall elevation. Not all falls from several hundred or perhaps a thousand feet, prove to be fatal, but on the average, the higher the elevation, the higher the severity. Gravity and concrete floors are very unforgiving. The adage "It isn't the fall that hurts you, it's the stopping" correctly identifies deceleration as the hazard. If fall protection does not properly distribute and control the rate of deceleration in falls from heights, injuries or death can result.

Falls from the Same Elevation

Slips and falls from the same elevation are often due to improper coefficient of friction, uneven working surfaces, misplaced objects in normal pathways, or visual effects.

Coefficient of friction should generally remain above 0.30. Liquids on smooth surfaces often reduce coefficient of friction, as does the area of footwear in contact with the floor. Ice drastically reduces coefficient of friction.

Fall Protection Systems

Fall protection is divided into passive and active systems. Passive systems do not require worker/victim positive action while active systems require a positive action, such as donning, attachment, and use. Nets are an example of passive fall protection. Extensive information on safety nets is given in a NSC Data Sheet (National Safety Council, Aug. 1983). An example of active fall protection is a harness and inertial reel with attachments. Arrangements for anchorage, attachment to the worker, absorption of shock, arresting, and use of a descent lifeline have been described as a "Fall Arresting System" (FAS) (Solowski, 1979).

Many approaches to fall protection have been used, with varying degrees of effectiveness. Barriers such as handrails, cables, and ropes have been used for years. Nets have been strung to catch personnel and dropped objects. Lanyards have been used to tie persons to structural support in case of fall. Many methods have been used to attach persons to the lanyards, such as belts and harnesses. The American National Standards Institute (ANSI) classifies belts and harnesses into four categories: body belts, chest harnesses, full-body harnesses, and suspension belts.

Ohio State University studied dynamic fall deceleration force effects in dogs back in 1946. Heart, kidney, liver, and brain damage resulted. Based upon these tests, 10 g (approximately 2,000 pounds) was recommended as a maximum deceleration for waist belts and 35 g (approximately 7,000 pounds) for a body harness (Damon, 1981). One g equals the force of gravity, caused by the mass attraction of the Earth.

Twin City Testing and Engineering Laboratory tested impact loads with a "standard" 6-foot lanyard that exceeded 2,600 lb.-ft.for a 130-pound weight. Loads for shock-absorbing lanyards reduced the shock as much as fourfold (DB Industries, 1983). The usual shock-absorbing lanyards are one-time-use items that expend energy by ripping or tearing fibers. Keep in mind that if the worker ties off and climbs upward, the total fall acceleration before running out of lanyard would be doubled. Injury thresholds depend on the time and points of force distribution on the body, and vary between individuals, but deceleration has been used as a limitation. An earlier OSHA Fall Protection Systems draft limited FAS deceleration to 10 g for belts and 25 g for harnesses (Sulowski, 1979).

The U.S. Air Force Systems Command's Human Systems Division performed a scientific study of physiological and subjective responses to being suspended in three types of fall protection harnesses (Orzech et al., 1987). Even without a fall impact, test subjects lasted roughly 1.5 minutes in body belts (difficulty in breathing), about 6 minutes in a belt/chest harness, and about 14 minutes in a full-body harness. The Air Force doctor in charge of the suspension tests concluded that body belts and body/chest belts should be banned. Findings from 1978 and 1985 studies warn of brain damage and eventual death if suspended vertically for 4–6 minutes, and irreversible protracted shock from pooling of blood in lower extremities. The implication is that without self-rescue or rapid assistance, a suspended worker in an inaccessible location is in serious danger, with extreme risk to individuals with cardiovascular disease.

Fall Control

A variety of controlled-descent devices have been developed that allow a person to descend to safety. These range from slides and chutes to tubular socks, cables guyed at angles, and controlled descent devices which use ropes and friction devices. This is an area where new technology continues to develop.

Another approach to minimizing fall risk is the use of climbing aids. These devices automatically support a climber via a permanently mounted cable or rail, using a harness and locking device, often inertia activated. Dr. Ellis points out that the anchorage support has to work and the connecting hardware to the body-support equipment must not fail either. He emphasizes that hardware such as D-rings, straps, and snap hooks are susceptible to misuse (Eisma, 1990). Standards require that critical weight-bearing items such as anchorage points and hardware be tested to twice the maximum expected force levels. Design safety factors are typically 4:1 or greater.

Nets are used as fall protection devices, particularly on elevated construction. Nets are often used as debris catchers and may double as fall protection devices if properly designed and maintained. "Personnel" nets have mesh sizes of 4 to 6 inches and are designed to catch a 350-pound object dropped 50 feet onto the net at various locations, the equivalent of 17,500 foot-pounds of force (Lahey, 1983). The U.S. Army Corps of Engineers has detailed net requirements in its Safety and Health Requirements document EM 385-1, April 1981 (National Safety Council, 1983).

FIRE SAFETY

Fire safety has been a common safety concern since the beginnings of industry, yet is still often poorly implemented despite many excellent codes and improvements in technology. There is a fine-line distinction between explosions and fires, and both are closely related. Explosions are treated in a separate section.

Fire and explosion losses represent tens of billions of dollars per year (NFPA Standard 6-10). Part of the fire-loss picture is confused by deliberate arson acts. Although arsons are declining, the National Fire Protection Association (NFPA) says that in 1989, 97,000 arson fires were reported, accounting for about $1.6 billion in property damage (about 20% of all structural fire losses.) Arson is the largest cause of property damage due to fire in the United States (*Professional Safety*, Jan. 1991). As much as 50 percent of arson may result from juvenile firesetters (McKinney, 1983). Yet, from a safety standpoint, arson fires may not be that different in effect from other fires and, as such, warrant engineering control where possible.

It is clear that fire deaths are a continuing problem, since that fatality rate has remained relatively constant at 10,000 plus or minus 2,000 since 1933, with 30 times that number of injuries (Jenkins, 1984). Fifteen percent of the total, or 1,500 deaths, are industrial fire fatalities. Fire injuries are about 30-fold higher (Jenkins, 1984). Smoke and toxic gases from fires are of particular concern because 80–90% of deaths are caused by smoke and gas rather that flames or heat (Bukowski, 1985). Computer modeling of fire spreadability is improving rapidly (Ben-Ali-Khoudja & Johnston, 1986).

Although 85% of fires are residential and 1,500 of the total deaths are at-tributable to smoking (Bottom Line Personal, 1989), there is still a large in-dustrial concern. For those interested in after-the-fact protection, the loss-fin-ancing approach to fire losses through insurance ignores the fact that 40% of all insured businesses never reopen following a fire (Dittman, 1988). Exclud-ing arson and insurance fraud, fire hazard is a critical business risk. Many insurers insist on their insureds meeting property protection requirements in order to be covered. The insurance companies probably account for the largest pool of fire safety professionals, followed by architectural engineering firms. Fire protection skill is a common need for most safety engineering practice, however specialization in this area often requires full-time commitment.

The etiology of fires is worth discussing. The well-touted "fire triangle" (fuel, oxygen, and ignition source) has been expanded to a "fire pyramid" adding free radicals to the three other necessities for fire.

Hazards

The main hazard in fires is toxic gaseous combustion products. The toxic by-products of a fuel, such as polyvinyl chloride (plastic) wire insulation which creates hydrochloric acid, can be a problem even if the majority of the by-product remains in the immediate site of the fire. This agrees with the fact that 80–90% of fire deaths are caused by smoke and gas, rather than flames. The National Bureau of Standards' Fire Research Center has developed computer models, such as NBS/COMPF-2 and NBS/FAST, to predict fire, smoke, and toxic gas throughout a building. Although carbon dioxide is a common toxic by-product, it by itself is not usually found at toxic levels. However, at 5%, by volume in air, it causes a deadly synergistic doubling of the effect of carbon monoxide. Hydrogen cyanide and carbon monoxide account for most smoke deaths (*Chemical & Engineering News*, May 20, 1985).

Additional firefighting hazards include flashover of superheated room gases and the destructive phenomena in large storage tanks know as boilover (White, 1987). After extended burning, either phenomena can envelop fire-fighters in a fireball. Thus personal protection for firefighters is of increasing concern. Structural failure is also a major fire effect concern for both the occupant and the firefighter. Fire hazard evaluation starts with building construction classification and building occupancy classification (light, ord-inary, and extra hazard). According to construction and occupancy, various levels of protection are mandated. In order to gain leverage from Pareto's

Critical Few principle (90% of problems are caused by 10% of the causes), safety engineers need to know that the NFPA reports the three top fire hazards on the job as trash, flammable liquid waste, and flammable liquid use (Stevens, 1985). In the plant environment, electrical fire hazard has been greatly reduced by the use of strict codes. Most of the fires that result are related to electrical connector failures (Stevens, 1985). Routine inspections and preventive maintenance should reduce connection hazards. NFPA 70B provides guidelines for electrical preventative maintenance programs. Cable tray fires are a serious fire problem because of their approximate 150,000 BTU/foot fuel loadings and because of insufficient fire stops at wall and floor penetrations (Kasper, 1982). This has great significance for heavily instrumented and heavily cabled operations, both structurally and for life safety where cable trays run within exit corridors.

Other than water, firefighting agents usually cause by-products of various toxicity during fire suppression, or are toxic before exposure to a fire. Some wits have noted that breathing enough water results in deadly toxic effect. Toxic effects of the relatively safe Halon™ agents must be considered, especially in multistoried buildings where the agent may concentrate at lower levels and where high percentage recirculation is required, such as in the space station or aircraft cabins.

Many industrial fires involve packing materials, and smoking or electrical equipment ignition sources. Preventive maintenance is identified in NFPA 70B as a systematic way to reduce electrical fire hazard. The use of plastics is rapidly expanding and often elevates fire risk.

Controls

As in most safety efforts, prevention is worth more than cure. Fire protection engineering dates from the early 1900s and has life, property, and business operations as its focus. Fire safety codes and life safety codes have been in existence in the United States for nearly 100 years. Various requirements for fire-worthy building construction and automatic sprinklers have shown to be of greatest protective value. According to 30-year NFPA statistics, sprinklers are 96.2% effective in controlling fires. Further, only one in 400 of the 8,000 annual fire deaths per year occur in sprinklered buildings and most of the time these deaths were the result of explosions or fires originating in unsprinklered parts of the buildings (Dittman, 1988). Obviously, sprinklers work and are cost effective.

Hazard control precedence first dictates removal and reduction of the fire hazards themselves. Noncombustible containers with self-extinguishing or fire control features for waste make good sense. Both safety and environmental constraints regarding flammable liquid disposal must be considered.

Building codes are another source of preventive fire protection standards that regulate construction and fire loads. In general, compared with NFPA standards, national building codes are directed more toward life safety than property issues. The UBC (Uniform Building Code), the SBC (Southern Building Code), the National Building Code (NBC), and the Basic Building Code (BBC) are examples of national building codes. Many states and municipalities invoke all or part of these codes. At least 38 different building codes exist in the United States (Ben-Ali-Khouda & Johnston, 1986).

Limiting the available amounts of flammable materials is one of the best ways to limit the effects of fire. Working amounts should be limited to a minimal daily supply. Similarly, incompatible materials should be segregated for fire protection. Oxidizers and reducing agents (fuels) should not be stored in close proximity to each other. For instance, fertilizer and oils should be separated.

Fires generally have predictable rates of development that suggest fire suppression actions. Sprinkler systems have been proven to be effective in overall fire protection usage, because of their reliable and early action. To be sure, tailoring of the sprinkler systems is needed to account for extra hazards, such as high stacks of flammable materials, areas where wet pipes would freeze, areas where water discharge is a last resort, and so on. Sprinkler systems have to be carefully designed by competent persons. It is important that sprinkler systems be correctly optimized in their design to meet desired performance. Loss of system integrity may occur due to explosion, malfunction of the system and/or operators, or natural disaster. A well-designed, but nonfunctional fire protection system is useless. Much programmatic emphasis is placed on avoiding closed supply valves, planned impairments, and premature shutdown of systems.

The simplest sprinklers (used the most) are wet pipe systems where the sprinkler heads hold back the water until their fire-sensitive (or damage-sensitive) elements open and then flow until valves are closed or water supply runs out. To avoid freezing, antifreeze, heating, or dry pipe systems that contain air under pressure are used. To allow alternative firefighting or shutdown of water-sensitive equipment, preaction systems and supervisory control systems have been developed. Combination systems and innovative sprinkler designs are common improvements.

In recognition that sprinklers are designed to control fires, and many times extinguish the fire, sprinkler activation signals to fire departments are recommended. Likewise, alarms to local personnel should be sounded automatically at every fire detection and sprinkler activation. Quick response is needed both for fire suppression and for fire evacuation.

Quite often, changes to the factors interfacing with a fire system cause it to fail. Changes may occur to construction and to occupancy. Breaches of fire separation and introduction of extra flammable materials is common where working environments are not routinely monitored.

Many otherwise capable systems have been negated by closed valves, predictable damage to supply systems, and premature removal during fires. Systems approaches that emphasize proper initial design, maintenance, operation, functional testing, and design updating better solve these problems.

Protective fire systems also include deluge units that can preclude otherwise likely ignitions of hazardous concentrations of flammable vapors, gases, and liquids. Other controls involve inerting and dilution ventilation. In inerting, an inert gas is pumped in to quench any fire. Dilution ventilation involves relatively large volumes of air to dilute the concentration of gas or vapor below the flammable limit.

Physical fire isolation barriers such as fire doors and fire walls must be maintained in operable status. Most fire equipment must be inspected at weekly to monthly intervals.

Where prevention and protection fail, fire suppression (firefighting) methods must be invoked. As with the *Titanic*, design protection alone without sufficient emergency lifeboats is foolish. Provision of portable extinguishers and other fire emergency response equipment must be considered by safety and fire protection engineers. But it is important to note that fire protection does not equal fire extinguishers any more than safety engineering equals hardhats and safety glasses. Planning, training, and testing of the emergency response operation is vital to success.

No extinguishing agent is universal, most having qualities better suited to one or more of the four classes (A, B, C, D) of combustibles. Class A is used for common combustibles (such as paper, wood, and most plastics); Class B is for flammable liquids (such as solvents, gasoline, and oils); Class C is for fires in or near live electrical circuits; and Class D is for fires in materials which continue to burn without exposure to air, such as magnesium and metallo-organic fires. Extinguisher symbols for each class are standardized both by a unique shape and color.

Some suppression agents, such as "Purple K," tie up the available free radicals (reactive ions formed in the combustion process) and so extinguish fires.

Fire science has also expanded known fire development processes to include various effects of container materials and configurations. Recently, computer-aided programs have gone beyond earlier deterministic (time-related) treatments to a probabilistic approach that uses modeling to evaluate fire spread with respect to containment or termination, fuel energy content, location of initial fire, and number of people. For instance, one sophisticated modeling technique generates Markov Chain Bayesian probabilities to predict area of fire involvement and a Goal Oriented Systems Approach (GOSA) barrier analysis to predict fire enclosure success without structural collapse (Ben-Ali-Khouda & Johnston, 1986).

Preparedness

OSHA Industry and Federal standards require both emergency action plans and fire prevention plans. Annual fire inspections are required that include means of egress, vertical fire integrity, alarm systems, fire suppression equipment, heating/ventilation systems, exit and emergency lighting, hazardous operations and materials, and elevators. Fire prevention plans include lists of fire hazards, lists of fire controls, control jobs and personnel responsible, emergency actions to be taken, a written training plan and a written maintenance plan. NFPA 101 life safety code requires essential personnel, floor plans/maps, emergency rescue and first aid, refuge/safe areas, and evacuation warden duties (29 CFR 1910.138).

Under OSHA, fire brigades used to fight other than incipient fires must have special training and equipment.

Many excellent references for fire protection exist in exhaustive detail in the National Fire Codes, Factory Mutual publications, and other sources discussed in the Safety Resource/Reference section. Knowlege and study in fire protection is a basic skill area which most safety engineers will need to expand well beyond this overview.

HAZARDOUS MATERIALS

Hazardous materials are common to most workplaces and, to some degree, to most homes. Hazardous is a relative term, just as toxicity is relative. For this book, hazardous materials possess a significant potential for injury or illness, given normal conditions of use. In general use, hazardous materials are commonly recognized by safety professionals as materials with flammability, explosivity, toxicity, radioactivity, corrosivity, reactivity, or physical effects that warrant special care and hazard controls. Examples of hazardous materials include flammable gases, liquids and solids, explosives and explosive dusts; poisons and highly toxic materials; radioactive agents; strong acids and bases; pyrophoric and highly oxidizing or reducing materials; and cryogenic liquids.

Recognition of hazardous materials, their hazardous and physical properties, incompatibilities, symptoms, emergency treatment, containment, storage, transportation and control methods are supposedly addressed in Material Safety Data Sheets (MSDSs). Darrell Mattheis of Organizational Resources Counselors has encouraged the concept of an electronic national MSDS repository to deal with the enormous glut of information existing in MSDSs (Mattheis, 1992). This subject is further discussed in the Safety Documentation section of the Introduction.

There are many sources listing hazardous materials, either singly or in combinations. A partial source list is provided in Table 4-1. Materials which have American Conference of Governmental Industrial Hygienists (ACGIH) Threshold Limit Values (TLVs) and/or OSHA Permissible Exposure Levels (PELs) may generally be considered as hazardous, especially if the underlying documentation is reviewed and the limits are due to reasons other than

TABLE 4-1 Hazardous Material References

ACGIH Threshold Limit Value Booklet and TLV Documentation

AIHA Hygienic Guide series

California Regulatory Notice Register (July 1990) Reproductive Toxins

Clean Air Act chemicals

Dangerous Properties of Industrial Materials, 7th ed. (Sax & Lewis)

DOT 49 CFR 172.10 hazardous materials table

Emergency Planning and Community Right-to-Know Act hazardous substances

EPA Carcinogen Assessment Office carcinogens lists

Federal Insecticide, Fungicide and Rodenticide Act chemicals list

Federal Water Pollution Control Act pollutants list

First Aid Manual for Chemical Accidents, 1980 (Lefevre)

Hazardous Chemicals Desk Reference, 1990, 2nd ed. (Sax & Lewis)

National Toxicology Program's Chemical Database and Chemical Data
   Compendium

NIOSH Registry of Toxic Effects (RTECS) toxic substance list

International Agency for Research on Cancer (IARC) lists

National Cancer Institute/National Toxicology Program bioassay program

National Library of Medicine Chemical Information System database

OSHA 29 CFR 1910, Subpart Z

RCRA 40 CFR 240-271 Designated Hazardous Materials

RSC Chemical Safety Data Sheets, Vol. 4—Toxic Chemicals, 1991

Safe Drinking Water Act chemicals

Solid Waste Disposal Act hazardous wastes

irritation. The Environmental Protection Agency (EPA) designates hazardous materials that, in addition to being hazardous in the occupational sense, are hazardous to animals and plant life. This text deals with occupationally hazardous materials.

Hazardous materials are controlled by (1) purchasing controls (use of buying standards, materials guidelines, material safety data sheets, and designated suppliers and users), (2) hazard communication programs to inform users of hazards, and (3) methods to avoid risk associated with hazardous materials. Thus hazardous materials controls avoid uninformed acquisition of hazardous materials and assure proper transportation, storage, use, and disposal requirements for hazardous materials. Hazard analyses should focus upon the inherent risks of hazardous materials and the risks of incompatibilities that may result from system failures.

Hazardous material controls typically involve containment, remote handling, rate of reaction control, exclusion of oxygen, exclusion of ignition sources, separation of reactive components and, finally, use of protective equipment. Precedence of controls should follow a standard order such as elimination through design, control by failsafe and failure tolerant design, control by barriers, containment or hardening techniques, warn through alarms and/or signs, training personnel to adapt by procedures, using protective devices, evacuation to safe locations, and justification of residual risk.

Hazardous materials are often encountered as wastes. For this text, waste treatment is discussed in the section on Sanitation. Hazardous materials are also covered in the Transportation Safety section of this book.

HAZARDOUS OPERATIONS

Recognition

Hazardous Operations (HazOps) include recognized hazards that equate to high risk, depending on the level of hazard control. Hazardous facilities are often classified, based on hazardous operations normally performed or hazardous equipment contained in the facility. Like hazardous materials, there is a lot of variation between organizations as to what constitutes a hazardous operation.

Generally, the following activities are typical of hazardous operations:

Flight testing
Rocket firings
Underwater diving

Explosives handling

Blasting

Punch press operation

Manual band saw operation

Asbestos removal

Batch and continuous loop chemical polymerization

Rubber calendering

Transfer of alkyl pyrophorics

Confined space operations

Firefighting

Parachuting

Logging

Log ripping

Building demolition

Excavation

Cupola tapping

Steel rolling

Underground mining

Oil refining

Oil drilling

Test flights

Crop dusting

Carrier night landings

Hazardous waste site remediation

Warehousing manual lifting

The safety of chemical processes is a common area of need, and thus provides a good example of a hazardous operation. Chemical processes embody many safety hazards involving toxic materials, high flammabilities, corrosive materials, high pressures, high heat, potential for structural failures, electrical and mechanical concerns, and cryogenic gas liquids.

The wide variety of chemical processes and the degree of business sophistication make this subject very broad and difficult to generalize. One thing is sure: there are few chemical processes that lack significant hazards. If those hazards are not properly identified, evaluated, and controlled, risks may be unduly high. Many of the topics discussed below are of chemical-process origin.

Detailed knowledge of the process, feedstock qualities, handling requirements, process conditions, reactivity hazards, and emergency responses are vital to hazard recognition. Appropriate safeguards and necessary personnel exposures must be considered.

System-safety techniques can and have been applied to chemical processes and have recently been mandated by an OSHA standard. For whatever reasons, OSHA failed to identify the technique as being "system safety." System safety can be applied retrospectively, although it is much more effective if practiced in design stages.

The hazard and operability study (HAZOP for short) has been used in various forms for more than 20 years, originating with a 1968 Purdue University chemical engineering workshop (Bullock et al., 1991). HAZOP is a systematic method developed for the chemical processing industry to deal with complex failure scenarios involving multiple independent events. HAZOP is not system safety per se but shares many commonalities. HAZOP was designed as a hazard evaluation group effort, involving teams of individuals with specialties such as operations, design, maintenance, industrial hygiene, safety, training, and management. Process equipment and instrumentation and piping drawings are a starting place in the HAZOP process. Freeman gives the following list of documents needed to complete a HAZOP review (Freeman, 1992):

- Updated Process and Instrument Diagrams* (PIDs)
- Operating instructions
- Plot plan with locations of equipment
- Plant site map
- Process flow diagrams*
- Process description*
- Past safety audit reports on the system
- Past accident/incident investigation reports
- Interlock descriptions and classification
- Interlock sensor set points* (temperature, pressure, level, etc.)
- Vessel capacities, design pressures, relief valve set points, relief capacities*
- Similar system HAZOP reports

"Guide words" and checklists are used to identify deviations from "nor-

*Essential for completion of HAZOP review.

mal" operating conditions. Bullock et al. (1991) lists example expansions of 21 guide words, adapted as follows:

No Flow	Wrong routing; Blockage; Incorrect slip plate; Incorrectly fitted N.R.V.; Burst pipe; Large leak; Equipment failure (C.V., isolation valve, pump, vessel, and so on); Incorrect pressure differential; Isolation error
Reverse Flow	Defective N.R.V.; Siphon effect; Incorrect differential pressure; Two-way flow; Emergency venting; Incorrect operation
More Flow	Increased pumping capacity; Increased suction pressure; Reduced delivery head; Greater fluid density; Exchanger tube leaks; Restriction orifice plates; Density or viscosity problems; Cross connection of systems; Control faults
Less Flow	Line restriction; Filter blockage; Defective pumps; Fouling of vessels; Valves; Restrictor or orifice plates deleted; Viscosity problems; Incorrect specification of process fluid
More Pressure	Surge problems; Leakage from interconnected high-pressure system; Gas breakthrough (inadequate venting); Defective isolation procedures for relief valves; Thermal overpressure; Positive displacement pumps; Failed open pressure control valves
Less Pressure	Generation of vacuum condition; Condensation; Gas dissolving in liquid; Restricted pump/compressor, suction line; Undetected leakage; Vessel drainage
More Temperature	Ambient conditions; Fouled or failed heat exchanger tubes; Fire situation; Cooling water failure; Defective control; Fired heater control failure; Internal fires; Reaction control failures
Less Temperature	Ambient conditions; Reducing pressure; Fouled or failed heat exchanger tubes; Loss of heating
More Viscosity	Incorrect material specification; Incorrect temperature
Less Viscosity	Incorrect material specification; Incorrect temperature
Composition Change	Leaking exchanger tubes or isolation valves; Phase change; Incorrect feedstock/specification; Inadequate quality control; Process control procedures

Contamination	Leaking exchanger tubes or isolation valves; Incorrect operation of system; Interconnected systems; Effect of corrosion; Wrong additives; Ingress of air; shutdown and startup conditions
Relief	Relief philosophy; Type of relief device and reliability; Relief valve discharge location; Pollution implications
Instrumentation	Control philosophy; Location of instruments; Response time; Set points of alarms and trips; Time available for operator intervention; Alarm and trip testing; Fire protection; Electronic trip/control Amplifiers; Panel arrangement and Location; Auto/manual facility & human error
Sampling	Sampling procedure; Time for analysis result; Calibration of automatic samplers/reliability; Accuracy of representative sample; Diagnosis of result
Corrosion/ Erosion	Cathodic protection arrangements; Internal/external Corrosion protection; Engineering specifications; Zinc embrittlement; Stress corrosion cracking (chlorides); Fluid velocities; Riser splash zones
Service Failure	Failure of dry instrument; Air/steam/nitrogen/cooling water/hydraulic power/electric power; Telecommunications; Heating and ventilation systems; Computers
Maintenance	Isolation; Drainage; Purging; Cleaning; Drying; Slip plates; Access; Rescue plan; Training; Pressure testing; Work permit system; Condition monitoring
Static	Earthing arrangements; Insulated vessels/equipment; Low conductance fluids; Splash filling of vessels; Insulated strainers & valve components; Dust generation and handling; Hoses
Spare Equipment	Installed/non-installed spares; Availability of spares; Modified specifications; Storage of spares; Catalog of spares
Safety	Fire & gas detection/alarm systems; Emergency shutdown arrangements; Firefighting response time; Emergency & major emergency training; Contingency plans; Exposure limits of process materials and methods of detection; Noise levels; Security arrangements; Knowledge of hazards of process materials; First aid/medical resources; Effluent disposal

Guide words should be expanded to suit the operations. Also, HAZOP forms should be tailored to the needs of the organization. Keep in mind that HAZOP, like other analytical tools, is not infallible and should be recognized as an aid to logical thinking and systematic coverage of hazards.

Consequences of process deviations are determined along with relative likelihood of occurrences. Finally, recommendations for improvements are made in terms of design, operations, maintenance and operations, and management (Gressel & Gideon, 1991). Thus, HAZOP uses an ordered inductive logic process and ranks hazard consequences using a matrix worksheet approach. Further HAZOP details such as forms and utilization are available in HAZOP materials and are beyond the scope of this book.

Howard Fawcett and William Wood state about safety and accident prevention in chemical operations, "It's not those hazardous chemicals that cause harm, it is their improper, uninformed use that causes harm. . . . Every accident is caused either by a failure to act upon what is known about the hazardous potential of a chemical or by ignorance of its properties" (Fawcett & Wood, 1982). This observation reinforces the concept given earlier that hazard effects vary depending upon exposure. Control of risks in any hazardous operation involves both the reduction of hazard and the reduction of exposure. General information about hazardous materials is contained in the preceding Hazardous Materials section.

INDUSTRIAL HYGIENE

As a profession, Industrial Hygiene (IH) is closely related to safety engineering with some overlap of functions. IH has been developing as a formal discipline over sixty years, focusing upon improving the occupational health of workers. Portions of other disciplines, such as toxicology, ergonomics and human factors, and health physics, have been integrated to assure worker health. (These other disciplines remain as specialties in their own right but may be an alternative job description for some industrial hygienists.) In fact, safety, as a unified concept, must include worker health as one of its major concerns.

Compared with safety engineers, Industrial Hygienists (also using the acronym "IHs") are scarce. In the absence of IH staff, many safety engineers are expected to shoulder IH responsibilities and to know when to obtain outside IH assistance. In many organizations, fortunately or unfortunately, IH functions have been subordinated to physicians. I prefer the alternative chosen by many companies of placing the IH function in an engineering organization. Nevertheless, IH practice requires close coordination with the medical function. It is becoming more common to see IH groups in a combined Safety, Health, and Environmental (SHE) group, with representation at high staff levels.

IH draws upon interdisciplinary resources from science and engineering backgrounds. In the past many IHs came from chemistry and physics backgrounds, but as the technology became easier to operate and read, the evaluation and interpretation sciences, such as applied statistics, epidemiology, information science, and decision theory, assumed stronger roles. The engineering disciplines bear heavily on control technologies in ventilation, noise and vibration, radiation, and ergonomics. The underlying objective in the practice of IH is determination of cause–effect relationships, with the control of causes central to effectiveness. IH has historically used the classic systems approach of Recognition, Evaluation, and Control. (Recently, "Anticipation" has been added to the R-E-C maxim.) Yet in a systems approach, anticipation must be used in the recognition function.

Technical functions within IH include a practical knowledge of:

- Air sampling (aerosols, dusts, fumes, mists)
- Biological hazards
- Cumulative Trauma Disorders (CTDs)
- Environmental controls
- Epidemiology
- Ergonomics and Human Factors (HF)
- Industrial Air Quality (IAQ)
- Noise
- Radiation protection (includes laser technology)
- Respiratory protection
- Thermal stress
- Toxicology
- Vibration

Air Sampling

Air sampling involves a considerable knowledge of new technologies in terms of sampling media, interferences, sampling protocols, analytical methods, evaluation of results, and both active and passive sampling methods. NIOSH "analytical methods" have standardized a portion of sampling, yet improvements continue. Use of certified labs is highly recommended.

Biological Hazards

Biological hazards have regained emphasis as a result of such things as Legionnaire's disease, DNA research, and IAQ concerns. Biological sampling and evaluation are quite specialized. The average IH will need specialized knowledge and skill to perform biological hazard evaluations.

Cumulative Trauma Disorders

Cumulative Trauma Disorders (CTDs) are an important subset of the ergonomics area within IH. CTDs are important enough to warrant separate treatment as a safety skill in the Ergonomics section.

Environmental Controls

An area of specialization within IH that deals with removal of hazardous pollutants from workplace atmospheres is environmental controls. IHs, for many years, have utilized filters, scrubbers, precipitators, mists, and dilution techniques to control concentrations below acceptable exposure levels. Environmentalists have been overlapping the IH activities in the workplace.

Epidemiology

Epidemiology is the study of the incidence, spread, and control of disease among large groups of individuals. As such, this science has application to industrial hygiene in the determination of cause–effect relationships in workplace populations. For example, epidemiology showed the relationship of vinyl chloride exposure to the incidence of a relatively rare cancer in populations exposed to vinyl chloride in precursor manufacturing and polymerization plants. Similarly, asbestos causation in industrial worker mesotheliomas was demonstrated by epidemiological methods.

Epidemiology involves the use of many statistical methods and studies. In general, these studies are either retrospective or prospective in design. Retrospective studies, such as case control studies, are hindered by varying documentation of variables and effects, since the collection of underlying data was not commonly defined before the design of the statistical test. On the other hand, prospective studies have the advantage of clear definition of data. However, prospective studies need to be carefully constructed to avoid biasing the results in the way that data are defined.

Ergonomics and Human Factors

The reader is referred to the separate discussion, Ergonomics.

Industrial Air Quality

Another area of IH that is receiving a lot of interest is that of Indoor Air Quality (IAQ). With the advent of "tight" (airtight) buildings and reduction of makeup air in the name of energy conservation, existing problems with indoor air quality are worsened. Sick Building Syndrome (SBS) is often signaled by a dramatic increase of complaints. One strong indicator is that workers get better during their time away from work, and get worse upon return. Pinpointing the causes for the complaints is generally a difficult task. Experience shows there may be off-gassing from paneling, carpet, paints, or biologic agents in the heating, ventilation and air conditioning (HVAC) systems. There may be in-building out-gassing processes such as blueprint processing, photo development, or aerosol fixing of seismic trace data. The complaints are frequently scattered by location and individuals, seemingly unrelated to ventilation patterns or placement. A composite of corrective actions are usually required, since single corrective measures are infrequently effective because they do not address the multiple causes of SBS. Rarely will air sampling show concentration levels in excess of exposure limits.

Thermal Stress

Exposures to both heat and cold extremes are common in industry. Thermal stress may cause effects on the unprotected or unacclimatized worker. When the body's normal temperature regulation systems are overcome, serious-to-life-threatening effects can result fairly quickly. The body's net heat depends upon several things—metabolic heat (depends on muscular activity), radiative heat gain or loss (depends mainly on clothing), convective heat gain or loss (depends on clothing and air movement), and evaporative heat loss (depends on sweat rate). Radiative and convective factors are three times as large as the evaporative factor.

The body's main mechanisms to combat heat loads are (1) sweat rate and (2) blood circulation to the skin. Sweat production is a dramatic reaction mechanism. Acclimatized workers can produce almost one kilogram (2.2 pounds) of sweat per hour for up to several hours before sweat gland fatigue or lack of liquid limits the sweat rate. A liter of sweat can cool about 2,500 BTU per hour. Dehydration in excess of three liters generally causes serious psychological disturbances due to salt-water imbalances. Blood circulation with skin acting as a radiator is similar to an automobile's cooling system. In addition to sweating, heart rate increases to radiate internal heat from the skin. Rates above 200 beats per minute may trigger heart failure.

Heat stress.   Heat stress (net gain of heat) occurs in both "hot-dry" and "warm-moist" industries. Heat strain is the body's total adjustive mechanism against heat stress: biochemical, physiological, and psychological. The normal body core temperature is near 37 degrees C (the familiar 98.6 °F). Two to three degrees Centigrade above or below this core temperature spells real trouble. Death may occur beyond seven to eight degrees either side of 37 °C.

Symptoms of heat stress include:

- Minor prickly heat rash from inflamed sweat glands
- Occasional muscular "heat cramps" from salt imbalance
- Common heat syncope from blood pooling in the extremities leading to dizziness or fainting (typically upon standing up)
- Heat exhaustion where circulation insufficiency and electrolyte imbalance result in a pulse rate more than 150 beats per minute (sweating increases, skin is cool and the face/neck becomes red, and the brain is being starved of blood)
- Dangerous heat stroke collapse of the core temperature regulatory system (diminished sweating, hot dry skin, and an oral temperature above 106 °F.)

IHs use heat stress indices such as WBGT (Wet Bulb Globe Temperature), CET (Corrected Effective Temperature) and HSI (Heat Stress Index) to quantitatively evaluate heat stress. There is still much controversy about which heat stress measurement is best. Modern heat stress instrumentation is improving the ability to rapidly evaluate industrial heat stress. In general, higher temperatures combined with higher humidities create dangerous heat stress. As an example, exposure to 95+ °F and greater than 80% relative humidity is considered very likely to cause heatstroke. Exposure to direct sunshine can increase some heat indices as much as 15 °F.

Control of heat stress. Control of heat stress typically involves these qualitative measures:

- Rest–work schedules, including acclimatization regimes
- Rotation of personnel
- Weight maintenance by drinking cool water (0.1% salt solution of about 1 tablespoon salt per fifteen quarts of water) to replace sweating loss
- Cooling in ~77°F "cool havens"
- Use of spot cooling ventilation
- Protective clothing, including cooling vests and caps
- Use of radiant heat protective shields

Stephen G. Gregg of the Gatorade Sports Science Institute Physiology Laboratory states that a 2% drop in body weight due to sweat fluid and electrolyte losses significantly alters temperature regulation and ability to perform work without dangerous core temperature increase (*Industrial Hygiene News*, Sept. 1992). Workers must rely on scales or schedules for drinking fluids to know when to replace fluids, because thirst is an inadequate indicator of sweat loss. Additionally, electrolyte and carbohydrate replenishment helps proper fluid absorption in the small intestine.

Of particular interest, is the Liquid-Cooled Garment (LCG) developed by NASA for astronaut use. It has been long recognized that from 25–40% of body heat can be transferred from the head. This condition, and facial sweating implications, demands that the head be cooled by a cap. Portable water-antifreeze cooling is designed to cool a maximum of about 1,400 BTUs per hour (the "heavy work" metabolic heat range). Battery-operated pumps circulate the coolant through a system of small tubes.

A less expensive heat control garment is the ice vest. Dry ice or frozen water/gels are kept from direct skin contact by insulation. Melting cools the worker's torso.

Cold stress.   Cold stress can cause loss of extremities through frostbite and death if body temperature drops about 8 degrees from normal. Chances of death are 33% if body temperature falls to 91°F (Kuhlman, 1988).

The body reacts to cold stress by shivering and shunting blood flow to core functions. Skin color changes and pain may give indication of cold strain. Extremities are most affected due to lessened circulation, contact with the cold objects, or exposure to the elements.

There are cold stress indices which guide safe exposures, similar to the heat stress indices.

Control of cold stress.   Footwear, handwear and headwear are very important to the control of heat loss and resulting injury. Hard-to-detect moisture levels in liners of personal protective clothing cause up to 27 times faster heat loss than dry liners. In the same way, steel toe caps and arches cause dangerous heat transfer (Heyl, 1986). Since upwards of 40% of body heat is lost through the head, headgear insulation against cold is vital.

Because the body generates water vapor during work, this water should be able to escape without letting moisture into the insulating materials. Gore-Tex™ is a relatively new fabric that allows such a one-way transfer of moisture. Wool has a natural ability to insulate when wet, only losing a relatively small percentage of its insulating ability compared with other natural wetted fibers. Other synthetic fibers such as polypropylene have the ability to wick away moisture from the skin. By combining several insulating mediums, a practical combination can take advantage of each material's strengths. For instance, an outer shell may be a GoreTex™-like fabric, with cotton backing, down fill and a chlorofiber-polypropylene layered inner fabric.

Arctic clothing must protect against wind, cold, and abrasion. The military and petroleum production companies have lots of valuable cold-weather experience. A vapor transmission layer, an insulating layer, and an outer protective layer are necessary to protect against extreme cold. Layering traps insulating air. Visibility of the clothing against snow is often necessary. Reflective strips aid in spotting people during darkness. An arctic "snorkel" parka provides multiple closures against the wind, a Wolverine-tipped tube that protects the face, and a snug fit below the waist to maximize protection. Well-insulated boots are key to keeping the feet from freezing. Where hard-hats are required, insulated liners and integrated masks are needed to accommodate snow goggles and keep out wind. Easily removable and cleanable liners are required for maintaining high insulation properties.

Frank Heyl makes the excellent point that circulation must be maintained in extreme cold situations and related the story of too-tight boot lacing which cost a worker toe amputations. Personal protective equipment must not create its own hazards, such as flammability, escape from wet clothing, or entanglement (Heyl, 1986).

Noise

Harmful noise is present in many workplaces. In recent years, numerous studies have tried somewhat unsuccessfully to determine a single value of noise that would result in loss of hearing for an individual. The 1970 Occupational Safety and Health Act used the 90 decibel (dB) A-weighted slow scale, per eight-hour workday American Conference of Governmental Hygienists (ACGIH) limit. Because dBs are logarithmic, and by choosing a 5-dB doubling, 95 dB for 4 hours is equal exposure to 90 dB for eight hours. This effective halving of the allowable exposure continues up to a maximum allowable 115 dB per 15 minutes. Impact noise of short duration is also capable of causing hearing loss. Many experts suggest that too many persons will be harmed by the 90-dB limit and that an 85-dB basis should be used with a 3-dB doubling rate.

Noise-measuring instrumentation has benefited from microprocessors and now automatically captures data as often as every few seconds, if desired. Dosimeters are miniturized and rugged enough to be worn by the worker, rather than having to take multiple sound level readings and correlate them to the worker's positions. Many noise specialists suggest that the latter is much more accurate, but I believe that dosimetry is improving rapidly and that the small errors tend to be insignificant if we look at trends. If the aim is legalistic compliance with a fixed limit, rather than preservation of worker hearing, high accuracy is of value.

To control noise, the spectrum of noise and its levels must be measured. For this reason, measurements in addition to noise levels and noise dose must be taken. Special octave band measurements and impact measurements are needed. Some degree of expertise is required to obtain good noise measurements due to noise physics and instrumentation characteristics.

Control. The control of noise depends on the frequency, noise transmission, reflection and reverberation, absorption, distance from the source, combined effect of noise, summed exposures, and many other factors. Noise is best controlled at the source, and certainly elimination by design is most effective, if affordable. Typically, noise sources are isolated, damped, cancelled with active noise techniques, filtered, muffled, or enclosed.

Enclosure, to be effective more than about 10-dB reduction, is usually very expensive because of the materials required and the tight construction. Sources that generate heat or involve materials feeding in and out cause problems with enclosures. Enclosures have been one of the few methods that usually work with large power presses.

Other examples of the control approach involve use of elastomeric or elastomer covered metal bins used to catch metal parts, various techniques to quiet air streams, use of lower rpms or quiet conveying methods, and many others.

A great debate arises about the use of hearing protectors (a form of personal protective equipment) instead of engineering controls. There is little doubt that 10 to 30 dB of improvement can be achieved by hearing protectors. Hearing protectors include insertion devices, such as foam and elastomeric plugs, and muffs that are heavier devices held around the entire ear. Plugs and muffs can be combined for a greater effect. Bone conduction limits the amount of protection that can be achieved without going to a full enclosure of the head.

Additional contention arises about how effective hearing protection is because of problems with fit and differences in field and laboratory noise level reductions. In my opinion, closed-cell foam plugs or custom molded plugs are the best choices for daily hearing protective devices. Bona fide comparison testing values should be reviewed when selecting the required array of hearing protection devices required by present standards.

Hearing loss.   Hearing testing is valuable for determining trends in hearing loss. There has been argument as to the value of audiometry with regard to when it is taken, attempts to cheat, the effects of transient medical conditions such as head colds, and concerns about what percentage is due to work exposures and what is related to aging, disease, medication, or off-the-job activity. A series of well-performed tests over several years, starting with an automated audiometry baseline is preferred. Other tests are being used, such as tympanometry, to evaluate hearing loss. Significant threshold shifts are a measure currently used by regulators to trigger required actions. Records should be tagged with exposure data and retained indefinitely for use in the inevitable lawsuits where hearing loss is claimed against a particular firm. Awards for severe hearing losses exceed $100,000 in federal hearing compensation situations. Hearing loss liability continues to be a costly proposition for many previous high-noise industries.

Vibration

Structure-borne (as opposed to airborne) vibration, may be characterized as segmented or whole-body vibration. Segmental vibration typically is transmitted through the hand to the arm, while whole-body vibration is transmitted through the feet or area of body which contacts a seat.

Effects that are associated with the whole-body resonance that occurs at 5 Hz include cardiopulmonary changes, fine muscle control (Dai-Hong Kam, 1981), hypertension, and genito-urinary effects (Gruber, 1976).

Vibration-induced segmental injuries occur in foundry operations, logging, paperboard stripping, stonecutting, mining, and a variety of manufacturing and assembly operations. Whole-body injuries occur to vehicle operators such as trucks, buses, agricultural machines, wheeled and tracked heavy equipment, and military vehicles, especially tanks.

A 1983 NIOSH vibration syndrome study reported 83% of exposed workers in foundries and 64% of shipyard workers had the syndrome, and 47% and 19%, respectively, had advanced (severe) vibration syndrome. Alarmingly, the average time for advanced symptoms for foundry workers was about two years (*NIOSH Current Intelligence Bulletin*, 1983). Gasoline-powered chain saw vibration syndrome was noted in 40% of lumberjacks studied (*NIOSH Current Intelligence Bulletin*, 1983).

Vibration may cause localized injuries such as vibration-induced white finger (VWF), previously known as Raynaud's phenomena. There have been at least 15 different hygiene vibration limits and none seems to establish a definitive dose–response relationship (Starck et al., 1990). The arm resonates both at 30–40 Hz and 5 Hz. Vibrations in the 40–125 Hz range are suspected of causing disease. Primary hand tool vibration occurs between 35–150 Hz (Radwin, et al., 1990), especially 40–50 Hz (Weaver, 1979). The amplitude of vibration is also an important factor in injury.

Jukka Starck et al. (1990) suggest that measurement of transmitted vibration (possibly proportional to compressive force in gripping task) is needed to understand the cause–effect relationship.

The frequencies involved depend upon rotational speeds, tool mass, static and dynamic forces of operation, amount of hand–source contact, direction of forces, arm muscle loading, and hand positioning.

Measurements of vibration are difficult with traditional vibration instrumentation, and they may be limited by task interference in close quarters and may not accommodate long task durations (Radwin et al., 1990).

Structural resonance usually occurs in trucks and buses between 1–20 Hz, with up to 2.5 m/sec^2 acceleration (0.25 g). Biodynamic strain calculations show a 5-Hz peak. Eyeball resonance occurs about 10 Hz (Gruber, 1976). Fatigue appears to be a predominant short-term effect.

Corrective measures for vibration syndrome include proper maintenance of tools, support of tools, introduction of vibration damping designs, redesign of tasks, introduction of work breaks or job rotation, and maintenance of warm body and hand temperatures, and dry conditions.

Respiratory Protection

Respiratory protection is important to industry due to its relatively common use to control otherwise harmful levels of toxic materials. It is a form of personal protective equipment (PPE) and is discussed in that section.

From an IH standpoint, respiratory protection represents a sophisticated control. The design of respiratory protection is a difficult engineering task because it must fit on highly variable face contours, be comfortable for long periods of wear, yet employ durable, cost-effective materials which last for years, be testable, be lightweight, offer effective protection against as many toxic challenges as possible. Design must also be compatible with other PPE, and not cause unacceptable physiological breathing resistance. Proper selection of respiratory protection requires a fair amount of expertise. Proper storage, use, and maintenance also require close supervision. The evaluation of respiratory protection effectiveness is a highly technical challenge as well. Fit testing and filter/cartridge testing are used to assure system effectiveness. Respiratory hazards include hazardous gases, aerosols, mists, dusts, plus biological and radiological hazards.

Respiratory protection includes air of "oxygen-supplying" devices which provide breathable air in oxygen-deficient atmospheres and air filtering devices which remove toxic materials from otherwise breathable air. An example of the first type of respiratory protection is an SCBA (self-contained breathing apparatus) which has a compressed air tank and a regulator to make air available at safe pressures. An example of the second type is a full-face respirator with dual cartridges.

Manufacturer's data should be consulted to select suitable respiratory protection, and a properly qualified specialist (i.e., a certified industrial hygienist) should review and approve selections according to the worst-case usage conditions. Components from different manufacturers are generally not interchangeable.

"Protection Factors" are assigned to respiratory protection ensembles according to the approximate magnitude of protection that should be provided. Positive pressure devices (i.e., the SCBA) have the highest protection factors (on the order of thousands). With proper training and backup systems, these devices are often used for rescues and firefighting within atmospheres immediately dangerous to life and health.

An in-depth explanation of respiratory protection and the mandated programs required by OSHA regulations is beyond the scope of this text. However, the potential for death or serious injury/illness by uninformed application of respiratory protection demands that safety professionals receive qualified assistance or become experts themselves.

LASER SAFETY

Light **A**mplification by **S**timulated **E**mission of **R**adiation (laser) is a technology which can have serious safety implications.. Insun Chang wrote an award-winning article that gives good information on lasers and laser safety. A laser involves a lasing medium that absorbs and reemits laser radiation, a pumping system that raises the medium lasing emissions to a higher state, using a resonant optical cavity. Laser light is produced in a narrow frequency band, is of one phase, and travels in one direction.

There are three kinds of lasers, depending on their type of lasing medium: solid, liquid, and gas lasers. A ruby crystal laser is an example of a solid laser and produces a powerful burst of light that can burn holes in metals. The neodymium yttrium aluminum garnet (Nd:YAG) crystal laser produces a continuous beam. The semiconductor gallium arsenide laser is a solid laser that is small enough to use in fiber optics signal transmission. Gas lasers, such as a carbon dioxide gas laser, can operate at higher power due to gas cooling effects and emit in a narrower frequency than solid lasers, offering advantages in measuring and communications. Liquid lasers, such as a methanol-rhodium dye laser, are either pulsed or continuous depending on whether they are powered by a flash tube or a gas laser. Importantly, they are the only type of laser that is tunable to different frequencies and are used in atomic- and molecular-level studies (Chang, 1986).

Laser technology is powerful and broadly applied in many industries today. Small amounts of beam divergence, high monochromaticity, and pulse/strobe capability allow high power with relatively low energy, use as a sensor in a narrow frequency, and strobing capability (*Man*, 1985). Lasers are used extensively in high accuracy measurement and alignment and for machining (drilling precision holes and and cutting or shaping patterns in many materials). Lasers are also used to weld hydroformed and drawn metals, thermoformed and molded plastics, and composites. Other uses include engraving and heat treating (especially semiconductor annealing), robot guidance, machine monitoring, production control, reading and mapping (contouring), high-speed photography, wind tunnel velocimetry (flow) measurements, medical surgery and welding, laser spectrography, reaction catalysis, communication channels, bar code reading, copying and facsimile machines, and Star Wars-type weaponry.

Hazards

With laser use increasing, laser injuries are increasing. In addition to eye and skin damage from beam exposure, other laser components have these related hazards: high voltage and capacitive charge, flash tube explosion, fire, cryogenic effects, potential X-radiation above 15kV, and toxic substance effects (Chang, 1986).

About 85% of laser-related injuries are to the eyes. Less than half of those surveyed reported use of laser protective goggles or glasses (*Professional Safety*, May 1991). Lasers can damage the eye in three ways: structurally, photochemically, and thermally (Chang, 1986). Continuous wave lasers can weld the retina and pulsed lasers can explode retinal tissue. Possible delayed effects from laser exposures on the retina membrane between the pigmented layer and the choroid coat of the eye have not been shown by monitoring laser eye accident victims (Sliney, 1986).

Although lasers were first developed in 1960, the first ANSI standard was issued in 1973 and has been revised regularly due to rapidly developing technology (Sliney, 1986). ANSI Standard Z136.1 uses four levels of lasers, classified by increasing power and effect. As modified by the Bureau of Radiological Health, Class I lasers (exempt lasers <0.001 milliwatt power) are not harmful under normal operating conditions. Class II (low-power lasers, <1 milliwatt power) are not harmful unless the viewer stares at the beam for a long time. Class IIIB lasers (medium-power lasers, <0.5 watts power) can cause injury either from the direct beam or reflected beams. Class IV (high-power lasers, >0.5 watts power) can damage both the eye or skin by direct and/or reflected beam exposure.

The Nd:YAG's infrared wavelength is invisible and causes many injuries. By contrast, the continuous wave (CW) argon gas laser is highly visible but still has dangerous energy. For instance, a reflected beam which contains only 1% of an initial 1-watt beam power still exceeds the allowable exposure by 10 times the limit. A repetitively pulsed laser requires that total-on-time pulse modification of the maximum permissible exposure be calculated for 400–1400 nm wavelengths. A pulse repetition frequency and correction power is applied to lower allowable limits. Q-switched and mode-locked lasers are considered more dangerous in general than CW lasers. Somewhat surprisingly, an increase in power in high-powered industrial lasers, such as the CO_2 gas laser, reduces the hazard of direct reflection at a distance because the laser tends to melt most flat reflective surfaces, producing a more diffuse specular reflection (Sliney, 1986).

Controls of Laser Hazards

Controls include those built into the devices and those implemented at a worksite. For the more dangerous Class III and Class IV lasers, a qualified Laser Safety Officer must determine a Nominal Hazard Zone (NHZ) where hazardous exposures are most likely. This involves specialized skill and knowledge. Often, NHZs will extend less than a couple hundred centimeters (Sliney, 1986).

Federal Drug Agency Center for Devices and Radiological Health (CDRH) system-safety requirements for the design and manufacture of licensed laser products include (Sliney, 1986):

- Interlocked or secured protective housings
- Remote connector interlocking of entrance doors
- Key switch operation
- An emission (laser operating) indicator
- A beam attenuator such as a mechanical shutter
- Specified warning labels
- Protective viewing optics such as a filter
- Operator controls located to minimize exposure
- Output measurement device within 20%*

*Required of medical laser products.

- Measurement calibration schedule*
- Laser aperture label*

A general practice should be to use remote video viewing of laser operations, especially alignment, and providing video filters in all optical viewing pathways (Sliney, 1986). Another general protective practice should be to keep background illumination high to reduce normal iris diameter.

Protective laser PPE eyewear is very frequency specific in its application. Therefore, "universal" laser goggles would be severely limited by frequency. Laser eyewear should be picked for the precise frequencies of use and the optical density measure of absorbing capability. A less desirable alternative to absorptive filters is reflective filters, which are usually very thin and thus susceptible to scratching, possibly allowing hazardous radiation penetration. Various plastics like polycarbonate are being used to reduce weight, deflect heat, and improve impact resistance. Absorptive layers may be used, which would be effective if scratch-resistant and of sufficient thickness. Laser-protective eyewear that meets ANSI Z136.1 should be clearly labeled with optical density and frequency protected.

The institute's factors used in determining appropriate laser eye protection include:

- Wavelength for laser output
- Potential for multi-wavelength operation
- Radiant exposure/irradiance
- Maximum permissible exposure (MPE)
- Optical density of eyewear at laser output wavelength
- Visible light transmission requirement
- Peripheral vision requirement
- Radiant exposure/irradiance and time factors to eyewear damage, including transient bleaching
- Need for prescription lenses
- Comfort and fit
- Degradation of absorbing media
- Strength of materials, especially shock resistance
- Capability of front surface to produce specular reflection (Tanner, 1990)

The laser equipment itself must also be properly maintained.

---

*Required of medical laser products.

A general safety checklist for lasers follows:

- Updated laser survey information (dates, names, positions)
- Operating procedure review (dates, names, positions)
- Hazard evaluations (hazard analyses, NHZ calculations)
- Hazard controls (interlocks, master switch, barriers, beam stops, personal protective equipment)
- Ingress/egress review
- Proper marking (tags, labels, signs)**
- Damaged filter use restrictions*
- Authorized operators/operator training*
- Attenuation to Class I levels*
- Not in use output control**
- Controlled use area*
- Emission indicators*
- Control of specular reflecting surfaces in beam paths*
- Restricted beam access/barriers*
- Enclosed indoor installations***
- Safety power shutoff switch***
- Remote firing/monitoring***
- Key switch master interlock***
- Laser safety information furnished****
- ACGIH and Maximum Permissible Exposure (MPE) limits not exceeded****

LIFE SAFETY

Life safety is a broad topic dealing with the safety of personnel in buildings and structures during emergencies, primarily fires and explosions, but also other situations including hazardous materials emergencies, bomb threats, earthquakes, and tornadoes and other severe weather, that would require rapid evacuation of personnel or movement to areas of safe haven. For this book,

---

   * Class III and IV Lasers.
  ** Class II thru IV Lasers.
 *** Class IV Lasers.
**** Class I thru IV Lasers.

life safety is taken to be the specialization within facility safety that deals with building design and the interface of operations that would tend to hinder egress in an emergency. There is overlap with emergency response, facility safety, and fire safety, all covered elsewhere in this book.

The National Fire Protection Association has detailed standards that address life safety in NFPA 101, "Life Safety Code." The Uniform Building Code and the Southern Building Code also address life safety. Life safety is normally associated with fire hazards, but the codes solve the problems raised by most other safety problems which demand egress.

The titles within the Life Safety Code can provide the following insights (which I have added in parentheses) to the overall life safety issue:

- Assemblies and educational occupancies (concentrations of people)
- Communications and extinguishing systems (alerts and control)
- Detection and correctional facilities (restricted egress)
- Fire protection features (failsafe protections)
- Health care occupancies (inability to egress)
- Industrial, storage and miscellaneous occupancies (increased hazard)
- Interior finishes, furnishings and decorations (flammability and smoke concerns)
- Means of egress (the keys to escape)
- Mercantile and business occupancies (moderate-to-high fire loadings)
- Residential occupancies (significant exposure times and fewer protective measures)

Earliest life safety committees (who produced the codes) were driven by "notable" fires that involved significant damage and loss of life. That code evolved into mandatory requirements that are based on hard-earned experience, representing perhaps the most significant fire safety improvement.

Hazards to Safety Egress

Many life safety features relate to the physical ability to egress, such as design of "exit ways," doors, stairs, ramps, smoke towers and other protective measures. Materials and type of fire-resistant construction have a large influence on smoke and fire generation and transmission.

Obviously, the ability to egress from danger depends on the ability to recognize the danger as well as the results from the danger. In the case of correctional facilities, locked doors would keep fit occupants from being able to egress. Similarly, egress from remote mine sites or an overhead crane cab would be more difficult than from more routine workplaces. Handicapped workers, including those temporarily handicapped by an emergency effect, would have more difficulty in egress than an unimpaired worker. Thus, when egress demands exceed worker capabilities (regardless of the norm), safety is in doubt.

Ability to see in dark conditions is addressed by emergency lighting life-safety requirements to assure safe egress in case of loss of power, which is often the case in fires of electrical origin. The National Electric Code (NEC) goes a step beyond requiring installation to requiring regular maintenance and written report of compliance. Additionally, building codes and local ordinances may also have emergency lighting requirements. New technologies improve battery, lamp life, and quality of light. "Smart" light microcircuit technology allows self-diagnosis and tying of all emergency lights to a central control station for automatic testing, status reporting, and anomaly alarming. This could dramatically reduce inspection, testing, and maintenance time, leading to improved availability, in addition to the generation of valuable compliance reports (Ruskouski, 1990).

Having working exit signs and exit lighting alone does not assure functionality in all situations. If the need to egress involves smoke, exit signs and exit lighting may be unusable. Placement of signs and lights above the bottom level of smoke could seriously degrade their utility. Placement of signs and lights close to the floor, where persons are directed to crawl in case of smoke, has been suggested. Self-illuminating signs, without need for batteries, such as tritium gas, are available. Luminescent tapes can last during egress, providing lighting is on prior to power loss. Chemical sticks can provide sufficient light for egress, provided storage conditions and storage life is adequate. Hand-powered dynamo lights can provide light if personnel are trained to use them while egressing.

High-hazard occupancies may limit ability to egress in cases of explosion or deflagration. Some hazardous locations' egress is normally constrained by physical characteristics of the layout. For such high-hazard occupancies, the life-safety code emphasizes the ability to suppress and control fire.

Facilitating Safe Egress

The life safety code uses computed risk factors to evaluate special high-hazard occupancies, an approach that could be applied to most high-hazard occupancies. Matrices, charts and graphs, and worksheets aid facility planners in complying with code requirements.

Computer modeling has seen increasing application in forecasting fire development and fire effects in terms of time and space. This is further discussed in the Fire Safety section.

Often there may be specialized equipment to facilitate egress during otherwise life-threatening conditions. Examples include water gel blankets, personal escape masks or hoods, or mouth bit respirators. Innovative escape methods have been devised such as a slide wire from drilling rig derricks, escape slides out of propellant manufacturing rooms, and socklike tubes that allow controlled descent from upper floors of buildings.

Just as protection has been added to make egress routes safer, protection can be added in the form of a safe haven. Safe havens are intended to be occupied during the course of an emergency. Differing hazards must be protected against in the safe havens. Examples include blast effects, fire, smoke, or other hazardous atmospheres. Safe havens may be used for high-rise buildings, remote parts of mines, ships, and other locations where egress to safety is not practical. Safe egress from elevations is discussed in the Fall Protection section.

Emergency rescue is an alterative to unassisted safe egress. In particularly hazardous situations, emergency rescue must be thoroughly planned and specialized equipment must be provided. Sufficient numbers of rescuers must be trained and qualified to perform the rescues. Emergency medical capabilities should reside in rescuer groups. Additional information is contained in the Emergency Response section of this book.

LOCKOUT/TAGOUT

Lockout/tagout refers to the control of hazardous energies, often in confined space, servicing, adjusting or maintenance activities involving machines that place personnel at elevated risk. These energies include mechanical motion, potential energy from pressure, springs, gravity, electrical or thermal energy. Of particular concern is inadvertent activation when personnel are in contact with the hazards.

NIOSH (National Institute for Occupational Safety and Health) research has found that packaging/wrapping equipment, printing presses, and conveyors are involved in most accidental injury/deaths involving hazardous energy. Seven percent of all workplace deaths and nearly 10 percent of serious accidents in many major industrial groups are associated with failure to properly restrain or deenergize equipment during maintenance (USDL, 1988). Machine operators suffer most injuries while cleaning or unjamming a familiar machine, but maintenance workers account for one third of injuries. Most injuries involve machines that are still running or have been accidentally activated. A sawmill industry study showed start-ups and unwanted movements to be involved about one-third of the time (Paques et al., 1989). Half of the time, no emergency shutoff was available.

Lockout devices are physical mechanisms used to effectively lock out or block, isolate or dissipate energy at points of control. Lockout devices should not be readily removed, bypassed, overriden or otherwise defeated. The taking of a physical action, such as removal of a fuse, although better than hanging a tag, is still susceptible to defeat by fuse replacement.

Tagout refers to a prominent warning device capable of being securely attached to an energy-isolating device. Tags should be distinctive and durable, and should identify the applier or authority having control of the procedure. It should clearly identify the equipment not to be operated and the time and date of application. Like a lock, only one person should set and remove the tag when the hazardous work is complete. Since tagout is essentially an administrative control, strong consideration of the more reliable (physical) lockout should be given in situations of potential death or serious injury. Combinations of both lockout and tagout may be used for greater protection.

The most common lockout device is a metal apparatus that accepts one or more locks and can act like the hasp of a lock to prevent a control operation. Whatever form, it should be durable and resistant to tampering. Some situations require the use of cables, plates, blind spools, and blanks to isolate hazards.

Control of lock removal is obviously a critical aspect of the lockout process. Many well-designed lockout systems require that each person install their own lock and that person be the only one to remove that lock. For that reason, locks should be clearly identified as to the worker they protect. In the case of large groups of people working closely, in the same location, one representative can be designated to maintain the lockout for the group. This practice does increase the likelihood of risk to the individual, as does having a master key for the supervisor to be able to remove locks. Many lockout systems require that locks be cut rather than unlocked, and then only as a last resort. Unauthorized lock removal should be cause for serious administrative action.

Lockout placement is another critical aspect of the process. If the lockout is placed where it does not positively control the energy, it produces a false sense of security. To guard against misplacement, lockout location and type must be preplanned and tested to assure functionality. For instance, locking out the low-voltage start switch without locking out the main power is unsatisfactory. Switches have been known to fail "closed," which would cause a start-up with restoration of power. The machine should be tested, under power, after lockout to be sure that it won't start.

Dissipation of energy is possible by bleeding off power from storage devices such as capacitors and by bleeding off trapped pressure such as in accumulators. Where energies cannot be completely dissipated, it may be possible to switch the system to a safer state.

Lockout/tagout systems need to be carefully integrated into operations. Clear, well-written plans need to describe both the actions required and the devices plus their location for use. Supervision should play an active role. Multiple hazards of differing types should all be controlled. Everyone expected to come into contact with the system should have appropriate lockout/tagout training. The key to whether or not a system is subject to lockout/tagout control should always be unacceptable risk.

Finally, lockout/tagout records should be kept for specified periods of time to spot adverse trends and to assist accident investigations if the system fails.

The OSHA general industry standard for lockout/tagout is titled "Control of Hazardous Energy Sources." Exemptions from this standard are provided for electric utilities, agricultural, construction, maritime, oil and gas drilling industries. Equipment adjustments and operations that do not involve access to hazardous points of operation are also exempted, as are cord-plug and hot-tapping operations.

OSHA regulations insist that employers who use tagout without lockout take additional steps such as removing a valve handle, opening an additional disconnect switch, or removing a fuse or printed circuit controller card. Specific training is required to highlight the limitations (and remaining hazard) of tagouts. Specific physical requirements for the tagout and its attachment are given, as well as content. Importantly, new or extensively repaired or modified equipment shall be designed with energy isolating devices after November 1, 1989. In terms of a performance standard, employers must develop an "energy-control" program. This program requires (1) detailed written procedures on shutdowns, equipment isolation, device application and removal, dissipation of stored energy, and verification of lockout-tagout conditions, and (2) different levels of training for both authorized and affected employees regarding recognition of hazards and methods of control. A system-safety type of requirement calls for an independent audit of the entire program, conducted annually and documented. Outside contractors must be informed of lockout-tagout procedures and prohibitions (Rekus, 1990).

MATERIALS HANDLING SAFETY

Materials handling is a basic function contained in most systems. Injuries and deaths regularly result from this function and are a focal point of many safety organizations. Materials handling may be done manually, or involve mechanical handling systems. Many ergonomic problems exist with manual materials handling, and many equipment design problems exist in mechanical materials handling. In general, manual materials handling involves a large number of less non-life-threatening injuries, while mechanical handling involves rarer but more severe injury or death potential due to the higher energies present.

The 1981 NIOSH "Work Practices for Manual Lifting" guide is being updated and will include asymmetry factors, such as twisting, and coupling factors relating to the utility of handles for grasping, as well as the previous horizontal and vertical distances to be moved and frequency of lifting. The action limit disappears in favor of one recommended weight limit. The standard does not yet consider effects of fatigue. Dr. Gary Herrin states that the revisions are likely to condemn about half the lifting jobs in industry, versus the previous 2% unsatisfactory lifting tasks and the 42% of lifting tasks which were above the action level (*Occupational Hazards,* June 1991).

Some 10–25% of all industrial accidents occur in materials handling. In terms of forklift operations, the most serious usually involves a tractor-trailer leaving the dock during forklift loading. Dock levelers and vehicle restraints offer the best control for this hazard (Eisma, 1990). Administrative controls such as requiring truckers to chock trailer wheels often fail because of lack of administrative span of control and rapid turnover. Moreover, the chocks themselves may be incapable of keeping the trailer in contact with the dock.

Although the ideal slope for a dock leveler is zero, 7–15% is acceptable, depending on the lift truck. This has often involved 6–8 foot levelers but may need to be 10–12 foot levelers if hand pallet jack operations are involved (Eisma, 1990).

A problem for dock-leveler and trailer-restraint systems is that of variable trailer height. Trailer beds now range from a height of 60 inches in container trailers to 36 inches in high-cube trailers. A common design solution is to have three different dock heights (40, 48, and 60 inches), depending upon traffic. Door openings and door heights also are increasing to 9 feet wide, and 10 feet high, respectively (Eisma, 1990).

Other safety problems exist, such as the levelness of the transition, tight maneuvering space, poor visibility of loads, other personnel working nearby, high carbon monoxide levels, high noise, adverse temperatures, and congestion due to multiple loading crews.

In addition to addressing the above concerns, operator training and qualification help, as do communication systems such as lights, horns, or speaker systems. It is important that training involve functional aspects and tests, to help convert book or classroom learning into skills.

Mechanical handling system accidents are often due to improper loading, rigging, controlling, or maintenance. Design safety factors may be introduced to allow for some of these conditions. Common safety factors for critical lift equipment are from 8:1 to 4:1—that is, a capability of lifting from four to eight times the rated load safely. Safety engineers should verify that equipment is being used as designed, within design parameters. Examples of these design parameters include allowable temperatures, other environmental conditions such as corrosive atmospheres, impact loading, radius and angle of lifts, required support, load sensing, and load limiting.

MINING SAFETY

Mining safety has a long history of progress. Losses have been great overall. The Bureau of Mines reported 7,000 deaths from explosions between 1818 and 1910. Several U.S. coal mine disasters killed from 100 to more than 360 persons each between 1907 and 1952. The Mining Enforcement and Safety Administration (MESA) was formed in 1977 to enforce safety and health legislative requirements, such as Public Law 95-164, Safety and Health Act of 1977. MESA was later renamed the Mine and Safety Health Administration (MSHA). MSHA parallels OSHA in its jurisdictions over coal, metal, and non-metal mines. MSHA has placed a strong emphasis upon enabling safety engineering in mining through training.

Mining includes underground mines, surface mines, and underwater mining (not discussed in this text). Surface mines do not have all of the hazards of underground mining but often have offsetting hazards relating to large draglines, large ore carriers, crushing equipment, and hazardous ore processing equipment.

Coal mine safety hazards include fires, dust and methane gas explosions, explosives mishaps, structural collapse, electrocution, falls, crushing, and strains. Deep mines have significant radiation and heat stress hazards, as well as inherent egress hazards. Health concerns include pneumonocosioses, radiation effects, toxic metal exposures, ergonomic stresses, noise, heat stress, cancer, and mutagenic effects. Workers, machines, and materials all have to coexist in close proximity.

Rescue is made difficult by deep shafts and long drifts in dark, often wet and cramped conditions. Ventilation generally has to be mechanically assisted, and loss of power may quickly produce hazardous conditions. Emergency equipment such as fire extinguishers and breathing apparatus must be properly located throughout the work areas, with attendant harsh conditions.

Dilution ventilation is used as a primary control for avoiding flammable atmospheres. Dust is commonly controlled with water sprays. Ignition sources often result from such things as electrical shorts and mechanical failures of conveyor bearings, not to mention smoking. Spontaneous ignition of coal is well recognized.

Ergonomic solutions to machine design that better meet the needs of miners represents a major opportunity to control injuries associated with unusual positions and high-force applications.

Rock drills and other equipment cause cumulative trauma injuries. Widespread use of compressed air with noisy compressors and other common noise-producing mining operations places miners at extreme hearing risk. Mufflers and hearing protection are the primary controls for noise.

Some mines have high levels of radiation, especially uranium mines. Tailings from ore processing also represent a major hazard to workers, families, and portions of the public that may be exposed. Radiation monitoring underlies most radiation protection programs in mining. Cumulative exposures are controlled below published limits, often by limiting work times.

Toxic mining exposures vary, including cyanides from leaching operations (if acidized), lead dust, and silica dust. Respiratory protection is important in those areas where dust concentrations may exceed the standards.

Mining safety is a specialty that currently has few comprehensive safety engineering texts. Several excellent, older individual Bureau of Mines publications addressing mining safety engineering aspects are listed in the References. Safety engineers working in mining should perform a literature search for other texts and consult MSHA for more recent training materials.

OFF-THE-JOB SAFETY

Off-the-Job (OTJ) safety has importance in protecting the worker in that home hazard exposure time is more than two-thirds of that time spent at work, workers injured or killed off-the-job are lost to the organization, and workers are lost to the organization when family members are seriously injured or killed off-the-job. Kuhlman reports that ratios of injuries off-the-job to on-the-job range from 18:1 to 28:1. OTJ fatality ratios are even higher at 35:1 (Kuhlman, 1986).

Home deaths in 1988 were about 23% of the 96,000 all accident deaths reported in the 1989 *Accident Facts*. Of these, the largest percent (almost 29%) were related to falls.

Hazards

Many workplace hazards occur in slightly different forms in off-the-job settings. The home is generally considered more hazardous than the workplace because of lack of hazard controls. A summary list of typical off-the-job hazards follows:

- Burns/scalding
- Boating and swimming
- Camping/hiking
- Cycling
- Electrocution
- Falls
- Fire/smoke inhalation
- Fireworks use
- Hobbies
- Jogging/running/walking
- Mowing
- Motor vehicle mishaps
- Natural disasters
- Noise exposure
- Poisoning
- Recreational shooting
- Recreational sports

When members of workers' families are seriously injured or killed, days are usually lost from work. In 1987, the U.S. Surgeon General reported that about 8,000 children were killed, 50,000 were permanently disabled, and about 12 million received medical treatment injuries. Motor vehicle deaths are the greatest cause of death to children below 1 year of age and are a serious factor in other child deaths. Some 50% of vehicle deaths occur to pedestrians. Of the 1,000 bicycle deaths annually, 600 are to children. Wearing of bicycle helmets reduces head injuries sevenfold and reduces brain damage eightfold (*USAA Aide*, Oct., 1989).

The second most common cause of death in children was drowning in swimming pools, bathtubs, buckets, and toilets. Of the 8,000 drowning fatalities a year in this country, more than half occur in swimming pools. The highest incidences are to children 15–19 and under 4 years of age (*Network*, Summer 1982).

About 250 cases of permanently disabling spinal injuries occur each year due to injuries typically suffered by 12–25 year-old males diving into water where they never dived before (Yoxall, 1982).

Choking, suffocation, or poisoning fatalities rate fourth in children younger than 4 years of age.

The most frequent cause of injury to children was burns from scalding and matches/lighters (*USAA Aide*, June 1989).

Workers' parents, often seniors over age 65, suffer 75% of fatal falls and 33% of home fire deaths (*USAA Aide*, June 1989).

More people die in smoking-material fires than from any other fire cause. Eighty percent of home heating fires and related deaths involve fixed-area and portable heaters or their vents. More than 20% of fire injuries in the home are due to cooking fires (*Safety & Health*, June 1988).

An estimated 77,000 persons were injured seriously enough to need emergency room treatment in 1981. From 50-64% of injuries were from rotary blade contact, depending on the type of mower. This injury often occurs when people try to clear clogs with the blade turning, pull the mower backward over the foot, or slip when mowing. Other opportunities occur when starting the mower or adjusting cutting height, or removing a cuttings bag while the blade is spinning. The 1982 Consumer Safety Product Commission (CPSC) mower safety standard, which requires a dead-man control to stop the blade in seconds and clutches to engage the blade, should reduce injuries as older equipment is replaced (Horner, 1982). Toys and other product safety hazards are treated in the Product Safety section of this book.

Controls

Safety engineering hazard controls should be applied to OTJ situations. Selected hazard controls include:

• Scalds from hot liquids can be controlled by resetting hot water heater thermostats to 125°F maximum, preferential use of rear stove burners, and use of approved breakable containers for hot liquids (such as Pyrex™ glass). Burns from matches and flammable liquids can be controlled by safe storage and training in proper uses of charcoal starter fluid, flammable aerosols, and gasoline. Leave fireworks handling to professionals.

• Adhere to U.S. Coast Guard (USCG) standards for safe boating, especially regarding the use of life preservers or personal flotation devices, a fire extinguisher, repair tools, flashlight, emergency signaling equipment (visual and aural), and nautical "rules of the road." Do not drink while driving, be cognizant and respect adverse weather, leave boating plans with others, and do not overload boats. Boating safety courses are available from organizations such as the Coast Guard Auxiliary, the Red Cross, and the U.S. Power Squadron. Swim in groups, use the buddy system, know how to cope with riptides, take swimming life-saving and survival training, fence pools, get swimming lessons if a nonswimmer, supervise children's swimming, have rescue equipment, get CPR training, and use approved flotation devices. Avoid diving into water, especially in unknown conditions.

• Backpackers and hikers need to be aware of exposure/survival techniques, the hazards of interaction with dangerous animals, drinking water hazards even in pristine mountain streams. The latest first aid advice for snakebite is to stay calm, safely identify the snake before it gets away, sit down in a safe place, place a lightly constricting band above the bite, and only if not fairly close to medical treatment cut small, shallow (¼" parallel to the limb) cuts through the fang marks. Suck out the poison (spit it out in case you have an ulcer), pinching and lifting the puncture area, if possible. Keep it up until reaching medical care, for up to 50% of the venom can be removed this way. If alone, walk at a normal pace. If help is available, keep victim calm and treat for shock. Remember that most snakebites are not fatal.

• Cyclists should wear protective headgear (such as those with ANSI or Snell Memorial Foundation seals of approval), increase visibility to motorists by wearing light-colored or reflective clothing, use fiberglass whips/flags and spinning bike reflectors. Other controls include proper selection and maintenance of off-the-road equipment, qualification of young riders for motorcycles and all-terrain vehicles. It is recommended to install a wheel retainer device on bicycles equipped with quick-release mechanisms (*Professional Safety*, Dec. 1989).

• Electrocution often results from contacting high-voltage supply lines with conductors, working home circuits without isolating power, and use of electrical appliances near sinks or tubs with water. Install ground fault circuit interrupter outlets above sinks and tubs.

• Falls in the home often result from loose objects on stairs and slippery surfaces such as with bathtubs and showers. Use toddler gates to protect stairs, keep cribs away from windows, use window safety latches, keep traffic areas clear, apply antislip materials to tubs, and teach the relax-and-roll fall technique.

• Fire hazard controls include the control of ignition sources such as gas water heaters in garages where gasoline is stored, elimination of smoking where flammable vapor concentrations can build up, use of smoke alarms and fire extinguishers, use of escape plan/drills, use of the stop/drop/roll technique, use of fire emergency plans, and knowledge of proper burn first aid (such as flushing burns with tepid or cool drinking water). Keep space heaters away from flammables and use with proper ventilation. Eliminate smoking in bed. Limit extension cord loads and do not route wires under carpets. Have wood-burning stoves and fireplace chimneys and vents cleaned whenever creosote builds up to ¼ inch (checked routinely with a mirror or using direct access) (*USAA Aide*, Fall 1985). Keep a clear area around fireplaces and use glass enclosures or tight screens to keep embers from igniting flooring materials. Install chimney flue screens to catch sparks and keep wildlife out. Hotel fires, especially in high-rise buildings, deserve preplanning the exit path (count doors and changes of directions, check if windows open and where the paths lead), plan to get up and investigate all possible fire emergencies, use stairs instead of elevators for emergencies, know how to barricade your room against smoke, draw water in tub for fire fighting in your safe haven, etc. (*National Safety News*, Jan. 1981). Ask for rooms below the fifth floor which are accessible to fire truck ladder rescues.

• The use of fireworks is a seasonal hazard that injures some 9,000 people a year in this country, with 15% of the injuries to the eyes (*Family Safety*, Summer 1982). Fireworks are a dangerous consumer product and should be left to professionals, especially not to unsupervised children because of the high risk of injury. Even the more benign burning sparklers cause serious burns to children wearing flammable clothing or by igniting their hair.

• Hobbies such as woodworking and machine shop work in the home represent a major source of injuries from rotating equipment mishaps (saws, drill presses, lathes, and so on.) Stripping finishes from furniture may cause dermatitis and serious inhalation hazards. Degreasing of metal parts often involve highly flammable and toxic solvents. Gardening and yard work involve high risk to the 4 in 1,000 people who are severely allergic to insect bites and stings. Anaphylactic shock can kill from a single bite or sting. Those with known allergies should wear medic alert bracelets/tags and have ready access to an emergency insect sting kit such as those with a premeasured injection of epinephrine and antihistamine tablets. Normal sting first aid includes removal of any stinger and application of a cold compress. Contact medical professionals with any sign of anaphylactic shock or abnormal reaction.

• Joggers, runners, and walkers need to exercise on the proper side of the road (nearest oncoming traffic), wear bright or reflective clothing, and protect against heat and cold stress. Proper running shoes, slow warmup, and use of buddy systems are also key to safe running. Identification should be carried in case of health emergency or other incapacitating accident. Special identification tags are made for lacing into shoes.

• Mowing should be done in hard-toed shoes, if not safety-toe shoes. Care should be given to foot placement during starting and whenever the blade is spinning. Personal protective equipment should include safety glasses and long pants to protect against thrown debris. Mow cross slope to avoid slips into the blade. Always use the grass catcher if removing it presents an open-discharge path. Allow a hot engine to cool before refuling and use an approved gas can, preferably with a filling spout, and handle gasoline in open areas. Disconnect the spark wire from the plug before doing maintenance on a mower.

• Air bag cars, antiskid brakes, antilacerating windshields, child-proof rear door locks, and collision-resistant cars are engineered safety features. Refer to make-and-model vehicle safety records reported in the nonprofit public service group Highway Loss Data Institute's Relative Injury Claim Frequency statistics. Some vehicles' injury rates are only 35% of the average while others are 60% above the average. Proper seat belt and child restraint device use can prevent many serious injuries and deaths. A properly used child restraint (safety seat) can reduce risk of fatality to a child or need for hospitalization following a vehicle crash by about 70% (Society for Automotive Engineering, 1992). Use approved car seat/restraint devices, lock inside door locks, keep kids inside car envelope, give street crossing training (teach "the driver won't see you," use "familiar landmarks for no-cross when hazards are closer than such and such landmark" technique) (*USAA Aide*, June 1989). When driving in fog, use low beams, keep wipers on, and roll down the side window to listen for traffic. It may be safest to pull off to a protected area.

• Response to natural disasters such as earthquakes, hurricanes, and tornadoes should be planned for in advance. Stock up, in advance, on emergency food and water supplies, first aid supplies, blankets, and battery- or hand-operated lights. Immediate protective actions probably will be limited to getting under a table or doorway, or if outside, running away from buildings, power lines, or stopping your car safely if on the road. Cutoff of natural gas, electricity, and water may be vital first steps after earthquakes. Avoid switching on lights after a quake. Use of safe havens such as inner closets and bathtubs during tornadoes and the abandoning of cars should save many lives. Safe actions in lightning storms, such as moving away from large, isolated trees, getting away from metal objects, getting out of the water, and staying close to the ground, are advised.

• Noise exposure is a consequence of many hobbies and recreations such as snowmobiling, shooting, racing, loud music concerts, and improper head-phone use. After avoidance, ear protection devices are recommended. Routine hearing evaluations are recommended whether or not ear protection is used, because hearing losses are insidious.

• Poisoning hazards such as household cleaning agents, pesticides, polishes, poisonous plants, liquor, medicines including prescriptions and over-the-counter remedies, astringents, and colognes should be eliminated or stored safely. Consider replacing highly toxic materials with less toxic substitutes. Label warnings should be followed. Emergency numbers for poison control centers should be posted, and emergency aids like vomiting agents (such as Ipecac™ and activated charcoal capsules) should be kept current and ready for directed use.

• Recreational shooting safety practices such as those of the National Rifle Association (NRA) should be known and followed. Hearing protection should be worn while shooting, especially target shooting. A three-year Georgia survey of falls from tree stands showed that 75% of hunter falls resulted in moderate to severe injuries. Further, 25% of those injured suffered permanent paralysis of arms or legs. Wearing of a safety harness was recommended (Schulz, 1991).

• Recreational sports, especially contact sports, have a high risk of sprains, strains (ligament injuries), and broken bones. Proper conditioning can reduce hamstring injuries. The use of rubber cleats instead of metal cleats will reduce ankle sprains. Breakaway bases help prevent fractures. First aid treatment for sprains, shoulder and neck injuries, and fractures should be provided. Appropriate protective goggles should be used by handball and racquetball players to reduce serious eye injuries.

OFFICE SAFETY

Hazards and Controls

Office safety may not contain many catastrophic hazards but it applies to a very large population of workers and there are significant losses to be controlled. The highest severity is usually associated with fires or building collapse, such as in an earthquake, and electrical shock. Once catastrophic risk is taken care of, the safety engineer should deal with the risks associated with other things such as falls, musculoskeletal strains, struck-by injuries, burns, cuts, VDT effects, and indoor air quality.

Fire. The high-rise building can aggravate the fire/evacuation situation.

Larger numbers of people can be at risk, beyond the normal reach of ground firefighting equipment, relatively inaccessible except from inside the building. Although buildings may be quite fire-resistant, many decorative materials that are fire hazards may be incorporated. As usual, the fire's smoke and gases kill more people than does the fire. Combustible fire loads must be evaluated and controlled prior to fire inception. Control of fire transmission, both vertical and horizontal, must be achieved. Sprinklers have the potential for reducing the losses dramatically. Smoke control is vital in high-rise fires.

Building collapse.   Building collapse—often triggered by earthquake, tornado, excessive snow accumulations, or fire effects— will severely delay all the normal emergency response assistance to victims. Location of victims, life support and rescue can be exceedingly difficult without proper preparation. Just as with fires, disaster planning for building collapse must encompass preplanning, training, and supplies to cope with the unusual and overwhelming demands on emergency services. Emergency communications and organization are covered in the Fire Safety section of this book.

Electrical shock.   Electrical shock has fatal potential due to the widespread use of 110-volt current and the associated vulnerability of the heart to 60-cycle current. Overloaded and damaged extension cords are a prime office hazard. Cords should be sized and routed to achieve protection. Grounding pins on plugs should be retained and accommodated by proper sockets. Preventative maintenance is a key to controlling electrical hazards. Electrical hazards are discussed in detail in the Electrical Safety section.

Falls.   Falls—including slips, trips and falls—often occur on steps because of improper coefficients of friction due to ice, water, or oily conditions. Other falls result from foreign objects in the normally clear walkways. Individual circumstances often determine the degree of injury sustained, but in general, falls account for the majority of serious office injuries.

Many preventative actions can help avoid falls, such as improving coefficients of friction by quick maintenance action, improved stair and handrail design, improved lighting, and improved work practices that avoid manual carrying of bulky loads on steps. For walking slips, a simple solution involves training personnel to slow down, take short steps, hold handrails, in addition to the proactive controls such as snow/ice removal and sanding/salting of walkways and stairs.

Strains.   Musculoskeletal injuries, often back strains, occur during lifting, picking up objects, and recovery from near falls. Moving of office supplies and furniture is a common source of strains. Also, many strains occur from chair seating mishaps.

Hazard controls involve using professionals for all office moves: limiting container weights below 35 pounds whenever possible for single-person lifts, or using more than one person; keeping the cubic size reasonable; providing better handles; using mechanical lifting devices; conditioning the individuals to better avoid strains; and, in general, training the workforce to use safe lifting techniques. Note that the standard advice to bend the knees and keep the back straight does not often make ergonomic sense. Getting help, avoiding twisting during a lift, and keeping the weight close to the body are much better advice.

Musculoskeletal injuries often involve the back. Correct lifting and materials handling practices, along with task and worker matching, and physical fitness will go a long way toward controlling these injuries. Surprisingly, many strain/fracture injuries occur from failure of chair springs when the office occupant habitually leans back. Beefed-up springs or preventative maintenance on critical parts may yield better results than the more difficult modification of behavior.

Struck by objects. Struck-by injuries in the office setting often relate to filing cabinets or doors being opened into walkways. Filing cabinets upset due to their high center of gravity (such as opening too many drawers simultaneously) and people run into drawers left open in normal pathways.

Serious burns may result from the ubiquitous coffee and hot water present in offices. Thermal shock failure of carafes can cause major burns that require skin grafts. The safety engineering solution would assure suitable and properly maintained carafes for the intended service. Another burn potential involves the clearing of jams from copying machines with unguarded/uninterlocked parts above 140 °F.

Ergonomic hazards. Ergonomic hazards in the office setting are normally associated with repetitive work. Productivity increases are driving computer utilization in the workplace. Unions have projected that more than half of the industrialized country's work force now use Video Display Terminals (VDTs) regularly. According to OSHA research, musculoskeletal disorders affect about 19 million workers annually and nearly one-half the nation's work force at some time during their working lifetime. Since 1986, Compliance Officer inspections of injury records has led to specially trained responders where ergonomic disorders are identified (Chapnik & Gross, 1987). There is no evidence that VDTs cause miscarriages or birth defects (*Professional Safety*, April 1985). Radiation levels are quite low and have not been shown to have adverse health effects.

With the use of computers there are basically two human factors problems: (1) poor physical arrangements of chairs, keyboards, source material, and VDT implementation that lead to musculoskeletal strain and cumulative trauma disorders, and (2) VDT monitoring problems which may aggravate visual problems. Often, screens or keyboards may be placed outside an optimal location for the worker, resulting in improper viewing distances and angles, or excessive strains on muscles, joints, and/or tendons. An injury award was granted in Canada by an appeals adjudicator for a "medically reasonable causation" of "trigger finger" (a nonrelaxing condition) (*Industrial Health & Safety News*, Aug. 1988). Glare, contrast problems, and screen design may aggravate vision problems or lead to fatigue-related symptoms. Labor unions have increasingly promoted VDT usage regulations, recognizing the increased stress factors including work-pacing systems (Chapnik & Gross, 1987).

VDT hazard controls include workplace surveys and redesigns (lighting, glare controls such as circular polarized screens), equipment and chair adjustments allowing body alignment, foot and wrist rests, document holders, exercise techniques, and work-rest regimens.

Emergency Response

Emergency response in the office setting is a factor that relates to egress, treatment of health emergencies, and rescue. Workers with known health concerns should be pre-identified, with plans to provide special care where it can be anticipated. The most common office health crises involve heart and respiratory problems. In case of heart attack and loss of breathing, immediate response must be given within the short 8-10 minute time frame to save lives and reduce the amount of injury to the victim. Most heart attack deaths occur before victims reach primary medical care.

Personnel qualified in CPR must be identified, coordinated with Emergency teams (E-teams,) and should have the support of companies for first-responder aid until health care professionals can take over. CPR qualifications more than justify the cost because both on-the-job and off-the-job benefits are provided.

OSHA

The acronym OSHA in the common safety vernacular refers both to the Occupational Safety and Health Act and the Occupational Safety and Health Administration. In this section, OSHA refers to the administration and OSH

Act or the Act means the other. The agency is a regulatory branch of the government and, as such, has not always been well regulated itself. Like many police forces, the controls over its powers take effect somewhat after the fact. Besides congressional overview, OSHA implementation is checked by the Occupational Safety and Health Review Committee established by the law, and other executive branch agencies such as the Office of Management and Budget and the Government Accounting Office. The OSH Act, effective April 28, 1971, also created the National Institute of Occupational Safety and Health (NIOSH) research organization. The Act was extended to the federal arena on February 26, 1980, by Executive Order 12196.

In my opinion, OSHA is enough of a political entity to depend on the executive branch politics as to how it does its job. For instance, during the more liberal Democratic presidential administrations, OSHA leaned toward liberal and labor union agendas. During the more conservative Republican presidential administrations, OSHA leaned toward conservative and business agendas. Arguments can be made on both sides of the issues that OSHA has or has not produced acceptable cost–benefit results. Many have addressed shortcomings in the OSHA program, including an excellent article by Benjamin Mintz (1989). At present, the pendulum is swinging toward more performance-related requirements and away from strict specification of methods.

Only in 1980 was the Fed-OSHA extended to the governmental sector, and then by presidential executive order. There was and still is some credence that all federal programs did not and are not able to practice what the government preaches in OSHA. Federal programs that were not historically served by Public Health Service safety and health professionals are catching up to their counterparts in general industry. In 1989 injury/illness rates for federal government employees were 5.5 per 100 workers, compared to 8.6 per 100 for the private sector (La Bar, 1991). However, the direct comparison fails to compare the types of hazards to which most workers are exposed. Professionalism among OSHA, reflected by recognized safety and health certifications, is on the rise, supported by job advancement criteria, pay, and better recognition.

Standards

As an agency, OSHA attempts to control hazards with the use of enforceable standards. In general, most OSHA standards are derived from industry consensus standards. Therefore, many OSHA standards contain specifications that were originally meant to be guidelines. More recent standards have been written from a performance viewpoint. Major (overlapping) subject areas of OSHA standards include:

- Administrative: Hazard communication; OSHA recordkeeping
- Electrical: Grounding; Insulating mats
- Explosives: Blasting agents; Explosives
- Fire: Fire doors; Fire protection; Flammable gases and dusts; Flammable liquids
- Hazardous Operations: Abrasive blasting; Abrasive grinding; Belt sanding; Boilers; Compressed gases and air; Elevators and man movers; Excavation; Fans; Pressure vessels; Process safety; Revolving drums; Spray finishing and painting; Storage areas; Welding and cutting; and Woodworking
- Industrial Hygiene: Air contamination; Blood-borne pathogens; Heat; Lighting; Noise; Radiation; Toxic materials; and Ventilation
- Life Safety: Aisles and passageways; Emergency action plans; Exits
- Machine Hazards: Band and cutoff saws; Guarding; Mechanical power presses; Power transmission; and Radial saws
- Materials Handling: Chains, cables, ropes and hooks; Cranes and hoists; Dockboards; and Forklift trucks
- Medical Services: First aid; and Physical exams
- Personnel Protection: Eye, face and head; Foot; Hand; Personal protective equipment; and Respiratory
- Sanitation: Drinking water; Lunch rooms; Toilets; Trash; Washing facilities
- Tools Hazards: Circular saws; Hand tools; Pneumatic tools; and Portable abrasive wheels
- Working Surfaces: Confined spaces; Floor condition; Floor openings and railings; Housekeeping; Ladders; Scaffolds and toeboards; and Stairs
- The subject of standards is further discussed in the Safety Documentation section of the Introduction.

The amount and quality of enforcement of OSHA standards depends upon a great many things, including budget set by Congress and proposals by the executive branch. Tough times have shrunk OSHA like many other agencies. Even before tough times, the Environmental Protection Agency (EPA) far outstripped OSHA in bureaucratic growth. In terms of budget, EPA is about 16 times larger than OSHA. The 1990 EPA budget *increase* was nearly double the entire OSHA budget (*Industrial Safety and Health News*, 1990). Ric Fulwilder suggests that EPA's restrictive powers over occupational exposures are clearly greater than OSHA, such as its Clean Air Act amendments regulating employee exposures, Resource Conservation and Recovery Act (RCRA) criminal penalties for knowingly allowing "serious bodily injury" exposure, Chemical Safety Audit Team evaluation of company accidental-release programs for high acute-toxicity materials, and Toxic Substance Control Act (TSCA) audits of OSHA injury and illness logs. Through potentially stronger leverage in the workplace via TSCA and Superfund Amendments and Reauthorization Act (SARA), EPA in many areas has become "the other OSHA" (Fulwiler & Hackman, 1990). Many OSHA and EPA inspectors will be cross-trained in the other agency's regulations to recognize possible violations and are directed to refer these to the appropriate agency (*3M*, Summer 1991).

Part of OSHA's resources come from fines collected for regulation violations. The growth rate in civil penalties is expected to average more than 50% per year for the two year period 1989–1990. A 1990 budget reconciliation bill authorized a sevenfold increase in civil penalties. This should generate $70 million in revenues in 1991 (*Occupational Safety & Health*, Dec. 1990). Since 1986, OSHA, through its violation-by-violation approach, has increased fines by proposing penalties as high as $10,000 per each cited violation. The highest to date was a proposed $7.3 million dollar assessment against USX Corporation. At present, the Budget Reconciliation Act allows a maximum penalty of $70,000 per willful violation (*Occupational Hazards*, June 1991).

A distinction should be made between Federal OSHA (Fed-OSHA for short—the acronym FOSHA never became too popular) and State OSHA organizations. The law provided for decentralization to the state level, provided guidelines were met. As of this writing, twenty three states and the two territories (Puerto Rico and the Virgin Islands) had full-blown state plans. Incidentally, this total has been fairly consistent since the 1970s. Additionally, two other states (Connecticut and New York) have plans which govern public employees only. Eight of the twenty three states produce their own standards. California, Michigan, Minnesota, and Washington provide most of the state standards leadership (Sheridan, 1990).

There continues a debate as to the proliferation of differing standards under state plans, which pose a compliance headache for multistate operations. Many question the "as effective as" requirement for state standards and also resent the "more effective than" requirements of the individual states. This is further exacerbated by smaller governmental entities such as counties and municipalities setting different and stricter requirements. The resultant cry is for precedence of the federal standards. The cumbersome process of OSHA standard revision and the inadequacies of minimal federal standards is further discussed in the Safety References chapter.

A recent twist to regulations has been the development of "mediated" or "negotiated" rulemaking where mutual parties consent to how to control hazards. Examples of this include the OSHA-EPA negotiated rulemaking on exposure to the chemical MDA and the Chrysler-UAW-OSHA repetitive motion agreement for the auto industry.

OSHA enforcement is mandated by the law and is also structured by agency internal emphasis programs, and qualification programs such as STAR. Recent years have seen expansion of voluntary consultation programs aimed at supporting the infrastructure as opposed to confrontation and litigation. In 1989, there were fifty-seven OSHA/State Consultation Projects (Department of Labor, 1989). The Cooperative Assessment Program (CAP) is an example of joint government, management, and labor team solutions to technical and feasible compliance methods in a specialized area (D. Burton, 1986). OSHA has long offered training courses which address compliance methods along with explaining the standards. The Department of Labor Fatal FAX Program is an informational program which describes fatalities and compliance findings associated with the approximate 1500 fatal accidents investigated each year, with an eye toward avoidance.

Companies and consultants have made lifelong careers out of preparing for and reacting to the OSHA inspection. This topic is considered to be too vast and too fluid for treatment in this text. There are many good sources of information, programs, and services such as The Bureau of National Affairs, Inc. (BNA), The Commerce Clearing House, Inc. (CCH), Business & Legal Reports, Inc. (BLR), and others which should be consulted by safety engineers regarding OSHA inspections. Jeff Vincoli (1988) wrote a good review in ASSE's *Professional Safety*.

One trend relating to enforcement has been the rising cost of fines and penalties for noncompliances. Record penalties (e.g., $7.3MM) (*Occupational Hazards*, Dec. 1989) continue to be proposed, and some sizeable, usually reduced, penalties remain after the inevitable contests have run their course. For five years, OSHA has followed the "Egregrious Policy" where companies are fined violation-by-violation. The recent Omnibus Budget Reconciliation Act raises some penalties sevenfold (Sheridan, 1991). Law firms continue to make profits on OSHA litigation. Litigations are often decided in the race to the courthouse where different Federal Courts of Appeal lean one way or the other. When a tie results, a coin-toss decision is made as to which court hears the case (and which way it probably will go) (Tyson, 1988). In addition to the monetary penalties, there have been criminal and civil convictions aimed at company executives and even safety engineers, usually anchored in negligent and willful actions which are contrary to the law and customary good practice.

OSHA inspectors have the right to entry, with proper credentials or a search warrant, have the right to inspect records, interview employees, observe work and the workplace, take samples, take pictures, make readings with instruments, etc., in order to evaluate the safety and health of the workplace.

Employers may perform side-by-side sampling and photography with the compliance officer, but compliance officers may have private discussions with workers. All the information to be used in contested cases will be available through the normal discovery process. On the whole, few cases are contested, only about 10% and some of those are settled out of court (Sheridan, 1991).

Although most cited OSHA standards change, it is interesting to note that fully 36% of FY 1990 citations were on the Hazard Communication standard and the balance of the top twenty citations were fairly evenly dispersed from 2% to 5% of the almost 122,000 total alleged violations (Sheridan, 1991).

In the twentieth anniversary year of OSHA, questions of reforming the Act still arise in Congress. Labor is pushing for worker rights to stop work if they believe unsafe conditions exist, worker right to contest OSHA citations and penalties believed too lenient, and mandatory joint management–labor committees at worksites (Sheridan, 1991). Many want the OSHA standard setting process to be quickened and increase the role of NIOSH. OSHA standards are in planning to address ergonomics, indoor air quality, bloodborne pathogens, and motor vehicle safety (La Bar, 1991).

An in-depth discussion of the OSH Act and its regulations is beyond the scope of this text. There are many available resources which track all the latest changes and interpretations, and provide assistance in how to comply with OSHA regulations. An example of a hard copy loose-leaf format regulatory service is the Bureau of National Affairs, Inc. (BNA) *Occupational Safety & Health Reporter*. It features expanded indexing, reference files of regulation and statute language, timely federal and state court rulings. Large databases often use optical discs to avoid tying up memory in hard discs or using numerous floppy discs. Similar services may be offered as on-line/on-demand computer services that can be accessed by a modem and a personal computer for a basic fee and user charges.

PERSONAL PROTECTIVE EQUIPMENT

Overall, only one in four injured workers were wearing Personal Protective Equipment (PPE) at the time of injury. An Eastern Research Group survey analysis (of 139,000 OSHA-200 forms) indicated that appropriate use of PPE could have prevented as much as 37% of occupational injuries and illnesses (La Bar, 1990). Although there are often better means to prevent injury and illness than PPE (per the hazard control hierarchy), PPE should be used as a backup to other means and when other means are not able to be used.

A 1980 BLS report states that skin diseases accounted for 43% of all occupational illnesses in the private sector. Hands are the most likely point of contact between chemicals and the skin.

Of the estimated one million U.S. job-related eye injuries a year, studies suggest that about 90% could be prevented with the use of proper eye protection. Corneal damage, which often results, is painful and slow healing due to lack of blood supply (Rosenwasser et al., 1985). A 1980 BLS study showed that 70% of eye injuries were from falling or flying objects, resulting in mostly minor eye injuries such as scratches to the eye. Sixty percent of the objects were small (like a pinhead) and were traveling faster than a hand-thrown object. Sixty percent of those injured wore no eye protection (Hall, 1987).

Regional studies of work clothing burns indicate that as many as 7,000 deaths occur annually. Burn recovery is painful and expensive (averaging 45 days at $1500+ per day) (Colver & Colver, 1991).

PPE is used to dissipate force and keep hazardous materials from contacting vulnerable parts of the body. In the past, many standards have included PPE requirements, ranging from very general to very specific. PPE protects all of the body, using different devices. Common PPE classifications and examples follow:

1. Head protection (hard hats, welding helmets)
2. Eye protection (safety glasses, goggles)
3. Face protection (face shields)
4. Ear protection (ear plugs)
5. Respiratory protection (respirators, SCBAs)
6. Arm protection (sleeves)
7. Hand protection (gloves, mitts)
8. Finger protection (cots)
9. Torso protection (aprons)
10. Leg protection (chaps)
11. Knee protection (kneeling pads)
12. Ankle protection (boots)
13. Foot protection (metatarsal shields)
14. Toe protection (safety toe shoes)

15. Body protection (suit ensembles, coveralls)

Often, PPE may have combined functions which provide protection in more than one area.

PPE, to be effective, must be properly selected, it has to be used properly, it has to be worn during hazardous exposures, and it has to be stored and maintained properly. The first consideration in selection should always be an assessment of the risk to be protected against.

An actual example regarding selection of foot protection shows the difficulties of selection. A wide variety of chemicals ranging from solvents to acids, were buried in a hazardous waste site in a thirty-year history. The site access was dusty; in trenches the footing was slippery; heavy equipment with suspended loads was involved. Some workers had to operate trucks, loaders, and dozers in protective boots. Others had to dig with shovels. Puncture hazards were uncovered in the digging. Changing seasons brought high heat, humidity, rains, and later sub-freezing cold and wind. Choice of one boot to handle all these challenges was impossible, but involved trade-offs. The initial selection was a high, black neoprene boot with a steel toe. Fresh cotton socks and personalized shock-resistant insole liners were provided. Boots were washed and dried in a drier every 2–3 days. Daily showers were required before changing in a clean room to street clothes. Yet several foot problems resulted in blisters, with one infection causing lost time. An alternative medium-height white neoprene "Shrimper" boot without a steel toe was offered, with no more foot infections or crushing injuries.

Other foot protection factors include electrical protection, conductivity of dangerous static charges in explosive environments, metatarsal hazard protection, fire protection and slipping hazards.

The part of the foot protected is important, as shown by a recent study where 40% of workers injured while wearing protective footgear were struck on an unprotected part of the foot (Smith, 1990).

Protective clothing probably had its origins in North Atlantic fishermen's oil-skin (linseed oil and varnish treated all-weather cotton) fabrics. War-driven impermeable suits were first made with natural rubber, but were replaced with synthetic rubbers such as Neoprene™ and butyl rubber. Later polyvinyl chloride, Teflon™ and Viton™ provided superior protective materials which are often used in producing gas-tight (Level A) suit ensembles (McDonald, 1989).

Many forms of PPE must protect against exposure to chemicals. Both penetration and permeation of protective materials are significant factors, due to the false sense of security which may be provided. PPE can trap chemicals in contact with the skin for extended periods of time along with skin irritation, and increased heat and humidity conditions. The National Safety Council (NSC) recommends that protective gloves have a minimum thirty-minute breakthrough (*National Safety News*, June 1988). However, Swedish testing of multi-component solutions yielded breakthroughs between one minute and >4 hours. Surprisingly, Viton™ breakthroughs ranged from 4–20 minutes (ketone components were suspected of contributing to the short break-throughs); neoprene and nitrile also had poor performances below 20 minutes. Butyl rubber and polyvinyl alcohol exceeded four hours, but all other materials were less than 30 minutes (Forsberg & Farriadis, 1986).

PPE reusability is often an issue raised by cost. If PPE can be adequately cleaned without degradation of PPE performance, its use can be extended. The alternative is disposables, which may not offer the same degree of fit, durability, or protection as a more expensive reusable product.

Cleaning of reusable PPE is difficult to assess in terms of retaining protection. Many permeability challenges will continue to migrate internally, despite external cleaning. Solvents which could extract the challenge can also degrade the desirable physical characteristics of PPE. A comparison test showed that aeration followed by washing with detergent was ineffective and that thermal decontamination (up to 100°C) was effective (Vahdat & Delaney, 1989). For this reason, the NSC Data Sheet for selection of chemically impervious gloves recommends that all gloves be discarded at the end of a work week if chemical contact may have occurred (*National Safety News*, June 1988).

Testing and qualification of PPE has been an issue for some time. Independent testing has been an objective, to assure efficacy of PPE. The Footwear Industries of America and the Safety Equipment Institute have performed independent testing. Third-party testing, such as by Underwriter's Laboratories or Factory Mutual, is less affected by internal financial pressures of the organization.

More particulars are given on certain areas of PPE below.

Head Protection

NSC data show that about 70,000 disabling head injuries occurred in the workplace in 1988. A 1980 BLS survey found that the typical head injury was caused by falling metal objects of at least eight pounds. An important feature of these injuries is that 91% of injuries to persons wearing hardhats (probably less than 25% of those injured) were struck other than on the top of the head (such as on the back, side, or front brim of the hat, or beneath it). Lateral protection has been developed but is heavier, larger, hotter, and more costly than standard hats (Minter, 1990).

Because of dangerous tipping forces created, hardhats should be essentially round without lips or ridges. However, strength and utility (such as rain diversion) can be improved by these features. Besides a smooth outside shell, the suspension is key to shock absorbing and distribution of force. Removing this suspension, improper assembly, or the placing of objects in between the shell and the suspension can defeat suspension effectiveness (Minter, 1990). Another common malpractice is drilling holes or painting hardhats. Both of these practices can weaken the shell.

Hardhats and their suspensions have definite life spans. Storage in direct sunlight or extremely hot vehicle cabs will reduce protection. Some manufacturers call for suspension replacement every year due to degradation from hair oil, dirt, and oxidation. Hats should be routinely inspected and replaced at regular intervals. Simple one-inch brim deformation testing of polypropylene hats is recommended. If struck a strong blow, hats should be replaced as a protective measure.

Eye Protection

The vast majority of protective eyewear is planar eyeglasses. Approved safety glasses have both impact-resistant lenses and sturdy frames which are designed to keep lenses from being driven into the eye. Polycarbonate lenses, due to their exceptional impact resistance and light weight compared with glass, are most popular. Surface-hardening treatments can increase their scratch resistance, but it is generally less than glass. Specialty tinting is available to absorb specific wavelengths, such as from welding. Laser eyewear is often made in goggle form to keep hazardous reflections from reaching the eye. Laser lenses are relatively narrow in the band of spectral protection.

Side shields and wraparound glasses provide increased protection for side-entering projectiles. However, they are not designed to protect against liquid splashes or other areas of the face. Chemical coverall goggles provide splash protection, but still may require safety glasses underneath if projectile hazards exist. Some goggles have antifogging design features. A full-face shield is designed to give fuller protection from low-energy foreign bodies approaching from the front hemisphere. Safety glasses and/or chemical goggles should also be worn under face shields if projectile or splash hazards exist.

Contact lenses cause a great deal of controversy as to whether they trap liquid and vapors, causing greater injury, or whether they protect the eye from a splash. The National Society to Prevent Blindness recommends that, except for situations in which ocular injury is a significant risk, individuals should be able to wear contacts at work. When the work environment entails exposure to chemicals, vapors, splashes, radiant or intense heat, molten metals, dusty atmospheres (Hall, 1987), or cryogenic liquids, contact use should be restricted. Workers should still wear safety glasses or goggles or faceshields, in addition to contacts. A current list of those wearing contacts should be maintained for first aid and medical providers. All responders should be trained in proper removal of both hard and soft contact lenses. Employers wearing contacts at work should be instructed to remove contacts immediately upon irritation, blurred vision, or pain.

Some plants take the position that eye protection should be enforced plantwide if any eye hazards exist.

Face Protection

Face protection usually involves the use of a face shield, although some helmets provide face protection. It is important to continue to wear eye protection beneath face protection if penetration or bypassing of face protection is possible.

Ear Protection

Ear protection involves external control of the noise such as with muffs or a acoustically treated helmet, or internal control such as with ear plugs. It is important to know the noise-reduction rating of ear protection, since performance varies widely. Practical experience of many has shown that expandable closed-cell foam ear plugs achieve a better fit than most flanged ear plugs. Many people have two different ear canal sizes.

Ear protection must not be allowed to deteriorate. External seals on muffs have a service life, depending upon the use conditions. Seals should be inspected and replaced as necessary. Internal plugs should be kept clean by washing and drying. Sanitary storage provisions are needed.

Custom-molded ear plugs have the advantage of personalization and better care than generic plugs. The net cost may be less than disposables, depending upon their service life.

Respiratory Protection

Respiratory protection as PPE is typically cartridge-type respirators and escape-type supplied air hoods which may be carried on the person. Respirators are limited to the amount of protection by the cartridge design, condition of the respirator system, and the fit achieved between the facepiece and the wearer. Cartridge selection and fit testing are two of the keys of the eleven-part OSHA respiratory protection requirements. Usage instruction, inspection, maintenance, and storage of respiratory protective devices are also very important.

Life-threatening exposures are not normally controlled by respirators which do not supply atmosphere at positive pressures. For routine PPE respirator use, secondary indications of overexposure are needed, such as biological assays of contaminant metabolites.

Hand Protection

Gloves can provide a barrier to chemicals, cut and abrasion resistance, vibration damping, cold and heat protection. But for a glove to be useful, it must remain functional in terms of comfort, dexterity, gripping, durability, and cost. Grip varies with the materials and the manufacturing process used. Fit and thickness are key to dexterity and feel which are necessary for many gloved tasks. Glove design patterns and methods of seaming affect both the fit and the penetration of chemical challenges.

Electrically insulating gloves and sleeves are available for high-voltage protection. These products are color-coded for different voltages. Where use is critical such as for linemen, deterioration must be avoided and testing programs must verify effectiveness.

Another alternative or adjunct to glove hand protection is the use of barrier creams. Three types of barrier creams are as follows (Lahey, 1988):

1. Vanishing Creams—soap and emollients which are slightly alkaline, easily removed with water. May slightly neutralize very mild acids.
2. Water Repellent Creams—examples include organic silicones, lanolin, petroleum jelly, and ethyl cellulose. Water-insoluble acids and alkalis are repelled. Note that alkalis emulsify the film rapidly.
3. Solvent Repellent Creams—examples include sodium alginates, sodium silicate, tragacanth, and methyl cellulose.

Specialty hand creams offer insect repellency, poison ivy, and UV protection.

Foot Protection

About 10% of lost-time injuries are due to foot injuries. Fifty-eight percent of foot injuries were caused by falling objects over 30 pounds, 16% by stepping on puncturing objects, 13% from objects rolling over the foot, and 8% due to exposure to chemicals or water. Only 16% of injuries involved employer-provided or partially reimbursed safety footwear. Over 70% of those injured received no instruction on foot protection (Kuhlman, 1989).

Body Protection

Limited-use (disposable) clothing's durability, impermeability, comfort, and protective features are rapidly improving. Lower cost makes them attractive either as a replacement for reusable clothing, where decontamination is not required; or for use as a disposable covering for reusable garments. KAYCEL was the first limited-use garment, developed for the medical industry in the 1950s. TYVEK™ was applied first as a grocery apron in the late 1960s. In the 1980s, TYVEK was laminated or coated with Saran™ for liquid and vapor protection (Kappler, 1988).

Wilcher (1987) lists the following suit-access features that can improve chemical holdout capabilities:

1. Sealed zippers and double zippers
2. Storm flaps
3. Zipper placements
4. Seam closures

Four major types of seams have been developed as follows:

1. Type I—Sewn seams (lowest protection)
2. Type II—Sewn and bound seams
3. Type III—No sewing required (NSR) seams
4. Type IV—Sewn & strapped seams (highest protection)

NSR seams utilize adhesives, heat sealing, or ultrasonic sealing to cover seams (Kappler, 1988).

The Industrial Safety Equipment Association (ISEA) recommends PPE garment base materials be reviewed for:

1. Chemical hold-out protection
2. Durability
3. Flammability
4. Ease of decontamination
5. Physical stability with temperature variance

Garment construction aspects for review include:

1. Seam type
2. Seam integrity
3. Type of entry or opening closure
4. Specialized design features
5. Sizing
6. Comfort
7. Cost
8. Packaging
9. Storability

The ANSI/ISEA 101-1985 standard calls for five different sizes, based upon minimum finished garment dimensions. This is an improvement over the three-sizes-fits-all approach (Wilcher, 1987).

Total encapsulating suits are a highly specialized PPE garment that is designed to protect against a wide range of life-threatening chemical challenges. ASTM F1001-86 specifies 15 challenge chemicals (acetone, acetonitrile, carbon disulfide, dichloromethane, diethylamine, dimethyl formamide, ethyl acetate, hexane, methanol, nitrobenzene, sodium hydroxide, sulfuric acid, tetrochloroethylene, tetrahydrofuran, and toluene) to be used in evaluating PPE clothing ensembles. Note that there is no universal protective material and that compromise is inevitable.

Some experts recommend that splash testing be considered more realistic than continuous contact testing (the present standard). Some solvent–barrier material combinations exhibit relatively constant breakthrough for both liquid and splash testing, but others show an intermediate increase in time between liquid and vapor breakthroughs. The latter response still would require liquid testing (Man et al., 1987).

Integration of PPE points out concerns about gaps between gloves and sleeves, boots and pants, and respiratory or ear protection and protective headwear (*Industrial Safety & Hygiene News*, July 1984).

Flame-retardant (FR) clothing is achieved by use of natural char-forming materials, and by synthetic materials with high melting/burning points, and by treating/coating of fabrics like cotton. Thermoplastics are not suitable because of melting and shrinking onto victim's skin. Examples of FR fabrics include aluminized fabrics, fiberglass, aramids, polybenzimidazole (PBI™), and preoxidized polyacrilonitrile (PAN). High-temperature-resistant fabrics include phosphorous chemical treatments like Proban™, Pyroset™, and Pyrovatex CP; SEF Modacrylic, aromatic polyamide-based Nomex™ fabrics, aramid-based Kevlar™, wool-rayon blends, and other specialty combinations (Morin et al., 1988).

PRESSURE SYSTEMS

Pressure systems have great potential for losses, due to the contained energy, contents, and secondary effects of failure. Pressure systems can fail in rupture (also known as burst) or leak modes. Fragments can be ejected if the materials fracture or if tears occur in closed patterns. If a seam or weak line of material gives way first, the more common "oil-canning" or splitting occurs with release of pressure without many fragments being released. Both rupture and leak modes release the material under pressure into the surrounding environment, so the effects of such releases must be evaluated. Flammability, toxicity, and physical effects occur. If a pressurized cryogenic liquid is released, massive cooling occurs as the liquid rapidly evaporates. Structures may be radically weakened by spills of cryogenic liquids. Depending upon the pressure and the nozzle effect that may result with a hole discharging gas, the noise resulting may easily exceed 140 decibels. Large amounts of water may be frozen out of the air from the leak's adiabatic cooling. Ice may block relief valves or overload piping support. Hazardous touch temperatures well below freezing may occur in metal parts such as valve handles.

The total energy release that results from a pressure system depends both upon the internal pressure and the area upon which the pressure is acting. Thus, the force associated with a differential pressure of 15 psia (pounds per square inch above atmosphere) over the inside surface of a 55-gallon drum is a considerable 1,200 pounds! Many such drums which were not designed for pressure service have failed with potentially catastrophic results. The effects on nearby people range from death to no injury, depending upon the mode of failure, debris path travel, result of the drum contents being released, position of the worker, etc. As stated earlier, an "accident" has occurred, regardless of the severity of the effect.

A great deal of the safety engineering that addresses pressure systems has to do with safety factors and safety margins. The first of these previously defined quantities, safety factor, refers to the amount of buffer or pad existing between the undesired event and the maximum planned event. It may be thought of as a percentage margin above nature's limit. The second, safety margin, refers to the margin above the system requirement. Margin of Safety = Safety Factor − 1. In aerospace practice, Margin of Safety = (Limit Load) × (Safety Factor)/Limit Load] − 1. Safety margins must always be zero or preferably a positive value.

Design considerations affect the system's ability to contain pressure, withstand pressure excursions, and failure mechanisms. Choice of materials, methods of construction and fabrication, methods of operation, maintenance, and emergency responses all affect the safety of pressurized system operation.

Many standards and guidelines exist for choice of appropriate materials, methods of connection, and control of pressure, depending upon the type of service. Pressure vessel codes and standards are given in the References section of this book.

One of the best lines of defense regarding failure of pressure systems is designing to leak-before-burst and careful selection of materials regarding service. Fracture growth is a major concern where many operating cycles are involved in the life cycle.

Pressure relief is a common protective control which works automatically, if properly designed and operated. Selection of pressure at which to relieve with respect to system pressure and other control regulation is given in codes and standards. In general, relief pressure is chosen no more than twenty percent more than system pressure and about 110% above the last upstream control pressure. Ability to discharge the full predicted flow of the system should be accommodated. Attention should be given to where the relieved pressure discharges with respect to habitability and potential damage to system components. The effect of human error should be considered with respect to blocking off of relief valves with shutoff valves immediately prior to the relief valve which are strictly for valve replacement when the system must remain pressurized. An additional concern occurs with flammable gas relief or venting, that of transiting through flammable mixtures in the relief piping. This is of concern in large-diameter piping such as the venting of vapors back to a flare when loading displaces air and vapors in barges.

It is a common requirement to proof test pressure systems to verify system design, such as 1.25 maximum operating pressure. Due to the potentially catastrophic effects of vessel rupture, many tests are done with water, due to its lack of compressibility. Where air testing is done because of impracticality of hydrostatic testing, tests should be done by remote means, keeping persons out of the danger arc (which may be sizeable).

A very common error with pressurized systems is the breaking open of a component such as a valve or coupling when pressure is still on the system. Depressurization may not be easily accomplished or verified due to system configuration (lack of pressure indicators and valves) or blockages which are the cause for maintenance. Line breaking may be hazardous due to trapped volumes. Besides stringent procedures for the relief and verification of pressure, personal protective equipment (eye and skin) should be required and used for line breaking.

PUBLIC SAFETY

Public safety is of concern to every safety professional in the context of this book as it affects the interface between industry and the general public. From a practical standpoint, customers, workers from other companies, consultants, government officials, visitors, and relatives of workers who may visit company properties warrant protection and consideration in safety planning. Others with a less apparent connection to the business, such as nearby residents or those who could be impacted by business activities or products should be considered. One area of public safety specialization is that of Product Safety. This book does not address "public safety" relating to governmental, especially municipal, services such as schools, transit, and public office building-based services. Transportation Safety can and often does involve public safety, but this is covered in its own section.

Safety engineers should review public safety hazards associated with their facilities and operations, as they affect the general public. Risk to the public should be evaluated and reduced to the lowest feasible level. Examples would include storage of hazardous amounts of highly toxic chemicals, such as liquid hydrogen sulphide, where a tank rupture could cause lethal concentrations of gas to drift past plant boundaries; and storage of an explosive material such as prilled ammonium nitrate fertilizer, which could cause dangerous overpressures and blast effects beyond plant boundaries.

Product safety is a complete discipline structured around the safe design and use of products. This area is driven in part by product liability legislation and in part by the large tort claim business associated with damage from defective products. Internationally, the European Economic Community (EEC) effected a products liability directive in 1988 which addresses three classes of defects: design defects, manufacturing defects, and warning/labeling defects. George Nassif stated that these directives will have the effect of greatly increasing the amount of product liability claims in the EEC (Nassif, 1991).

The Consumer Product Safety Act (CSPA) has been in effect since 1972. It attempts to make companies more responsible for safe design of products, transferring regulatory responsibilities from several existing safety-related acts, but exempts several areas of consumer products covered by other acts or powerful lobbies. The purposes of the act are as follows:

1. To protect the public against unreasonable risks of injury associated with consumer products
2. To assist consumers in evaluating comparative safety of consumer products
3. To develop uniform safety standards and minimize conflicting state and local regulations.
4. To provide research and investigation into causes and prevention of product-related deaths, illnesses, and injuries.

There are product safety implications where subcontractors supply parts or assemblies for integration into a company's product. These suppliers' and vendors' product safety must be monitored and integrated into the total company effort. Business functions other than manufacturing are introduced through the definition of those who sell products: wholesalers, middlemen, distributors, and retailers.

A bipartisan Consumer Product Safety Commission (CPSC) is supported by a Product Safety Council (government, industry, consumer organizations, and community leaders) in development of standards, banning unsafe products, petitioning court seizure of imminently hazardous products, testing and labeling and certification of products, and requiring of repair, recall, or rebate in cases of noncompliance.

Both civil and criminal penalties are provided for violations of the Act. Private actions are encouraged via U.S. District Courts. States and political subdivisions are required to provide equal or stricter standards than the CPSC. However, this act must defer to the Occupational Safety and Health Act, the Atomic Energy Act, the Clean Air Act, and the Public Health Service Act.

Mandatory reporting to the CPSC of "defects" (any condition or characteristic of a product whether by design or by manufacturing error, which represents a risk of injury to a person when used in a foreseeable manner, other than a risk-producing characteristic obvious to the user and essential to the utility of the product) that represent "substantial product hazard" has lagged many expectations.

A National Electronics Injury Surveillance System (NEISS) was formed to collect and process product injury data. NEISS estimated that, in 1985, over 837,000 emergency room treatments were associated with stairs, followed by over 514,000 for bicycles. Upward of 50,000 emergency room-treated injuries occurred with powered lawnmowers in 1989. This statistic shows that injuries from consumer products are significant.

Product safety must be managed in a preventative, proactive mode like other safety. Safety should be involved in setting specifications, design review, safety input to reliability, maintainability and quality programs, labelling and instruction preparation, complaint review, product safety monitoring and reporting, product accident investigation, and review of design/manufacturing modification.

Sterling Hight (1981) gave this excellent checklist in 1981 for correction of product safety deficiencies:

1. Management participation
2. Hiring, selection and training key personnel
3. Product design and design review
4. Documentation and change control
5. Purchasing controls
6. Production controls
7. Distribution controls
8. Product quality control
9. Metrology and standards
10. Advertising, instructional and hazardous material literature
11. Sales control
12. Consumer or customer services
13. Accident, incident and customer complaint investigation
14. Product emergency procedures
15. Product recall or modification program

System safety analyses have use in identifying critical parts or functions capable of resulting in serious bodily injury, property damage, business interruption or incidental and/or consequential damages. Failure modes and effects analysis and reliability analyses may also identify critical parts.

Kaz Darzinskis (1990) stated in a reader's comment that "Of the [product sellers'] control that does exist, most comes from product design factors rather than from signs or instructions" (Darzinskis, 1990). Product design corrections should be made using the usual engineering precedences (eliminate, substitute, enclose/minimize energy, enclose/guard, safety devices, protective equipment, train/educate, and finally warn/use administrative procedures).

After a product reaches its useful life, or after major technological improvements have improved latter products, redesign should be considered as a method of reducing product liability. A 1975 National Builders Association study of power press age in product liability cases revealed that about 50% of presses were over 20 years old, 25% were over 30 years old, and 14% were over 40 years old (Adams, 1976). Since technology is changing at an increasing rate, redesign may be a cost-effective alternative to defending numerous product liability cases with "state-of-the-art at the time" defenses.

Joseph P. Ryan's (1988) excellent list of critical factors relating to safe use of products is adapted below:

1. Product Design Objectives
   A. *Consumer Use*: safety, functional, efficient, reliable, attractive, utility, life
   B. *Constraints*: cost, market pricing, economic life, sales, competition
2. Product Hazards
   A. *Energy Level*: pressure, force, temperature, explosive, flammability, mass, stability
   B. *Consumer Injury*: cutting, crushing, pinching, burning, abrasion, sharpness, tearing
   C. *Consumer Health*: hygiene, toxicity, radiation
   D. *Surface*: roughness, irregularity, sharpness, hardness, softness, slipperiness, stiffness, heat transfer
   E. *Geometry*: grasping, holding, bulk, center of gravity
   F. *Motion*: dynamics, rotation, reciprocation, intermittency
   G. *Safeguards*: guards, presence-sensing, emergency-stop, failsafes, protective equipment or clothing
   H. *Warnings/Instructions*: labels, signs, instructions, operator manuals, assembly/disassembly, clarity
3. Consumer Identification
   A. *Skill Required*: minimum, mechanical, chemical, electrical, prior experience, education
   B. *Mental*: age, analysis, alertness, impairments, maturity
   C. *Physical*: strength, gender, dexterity, handedness, height, balance, agility, stamina
   D. *Cognitive Processes*: perception, language, long/short term memory, risk-taking, comprehension, decisions
   E. *Psychological*: behavior, emotions, impulses, panic responses, deliberation, caution

4. Alternative Designs

   A. *Trade-offs*: product revision, competition, consumer safety

   B. *Safeguards*: consumer, costs, practicality, advanced technology

   C. *Constraints*: competition, benefit–cost analysis, pricing

   D. Discontinue or revise

5. Operation Manuals

   A. *Safety*: prominence, page locations, graphics, pictograms, safety advice, 800 numbers

   B. *Instructions*: safe use, pictograms, simple assembly steps, common tools, strength requirements, assistance requirements, cleaning, maintenance, storage

   C. *Warnings*: cautionary instructions, labels, signs, conspicuousness, durability, attention/alert emphasis, pictograms

   D. *Spare Parts*: illustrations, identification, ordering

   E. User Post-Purchase Safety Notification List

Third-party certification of compliance with standards, codes, and specifications is of major assistance in manufacturer's product assurance programs. The full variety of product safety certification approaches was discussed by Gerald Lingenfelter (1988).

D.L. Krause (1985) suggests that preassembling complete and accurate documentation is of great assistance in defense of product liability cases, and has been adapted below:

1. Quality-control methods used in manufacturing processes

2. Safety-related reasoning used to finalize the product design

3. Recall history, if any

4. Reason for presence or absence of any recall

5. Product manufacturing dates

6. Product identification (make, model, size, capacity)

7. Patent information, if applicable

8. Warnings, instructions, cautions in effect

9. Warranties and disclaimers

10. Provisions for service and repair

11. Industry standards applicable

12. Government regulations applicable

13. Improvements made in the meantime

14. Product safety profile

A product-safety profile is an anticipatory tool for identifying product hazards, and should contain information such as was adapted from a *Travelers Product Liability Newsletter* (1978) product hazard evaluation list below:

1. Potential loss-producing modes and/or events
2. Controls existing or needed to minimize hazardous effects
3. Clear and concise definition of the product
4. Probabilities of failures
5. Descriptions of product life, product uses, and servicing
6. Description of projected misuse
7. Description of failure prediction and symptoms
8. Description of failure effects including criticality
9. Description of environmental effects during life cycle

ROBOTICS SAFETY

The increasing use of robotic devices in industries involves increasing risk to workers. George Pearson's (1989) list of typical robotic applications has been expanded below:

1. Materials handling
2. Welding
3. Machining
4. Spray painting
5. Assembly
6. Test and inspection
7. High-pressure water blasting
8. Asbestos removal
9. Remote underwater operations
10. Sandblasting
11. Bomb handling
12. High-toxicity laboratory operations
13. Remote sensing operations

George's suggestion that robots be used to test and evaluate confined spaces is an excellent one. The advent of expert systems and automation improvements in technology should allow for a wide variety of expansion in robotics in the near future.

A typical industrial robotic system is shown in Figure 4-5.

The Robotics Industries Association defines "robot" as a "reprogrammable, multi-functional manipulator designed to move material, parts, tools, or specialized devices through variable programmed motions for the performance of a variety of tasks." The key word is "reprogrammable." Robots use software programs to achieve robotic performance. Industrial robots have been evolving into far more sophisticated devices than the single-arm connected to a stationary base. The need to handle highly toxic or dangerous materials, such as bombs, has led to many robotic applications. Similarly, robotic arms, known as Remote Manipulating Systems (RMS) are routinely used in space to perform demanding work in a harsh environment.

Due to the power of some robotic machines, people have been injured and killed by accidental contact between robots and humans (J. Burton, 1988). One Japanese repairman killed was inside an area marked "off limits," working on a machine still powered up. Touching the wrong transistors caused the machine to pin the worker and fatally stab him. Paul Scheel indicated that five

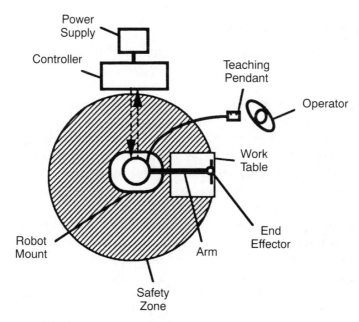

FIGURE 4-5 A Typical Robotic System

unpublished fatalities have occurred in Japan, and at least two in the U.S. (Scheel, 1993). Ironically, one of the robotics deaths involved the operator being pinned against a safety pole.

In defense of such robots, they are not programmed with an intellect like Arthur C. Clarke's computers HAL 9000 or SAL 9000 (Clarke, 1968), so people must take responsibility for staying out of a robot's way.

Robotics Hazards

The following is a list of typical hazards associated with robotics:

1. Impact/struck/shear and pinch due to robotic equipment motion
2. Impact from dropped robotic loads (gravity)
3. Impact from ejected robotic loads (rotational or linear acceleration other than gravity)
4. Electrical shock
5. End-effector effects (welding arcs, lasers, high-pressure jets, spray paint streams)
6. Improper information from remote sensing applications

Discussion of Hazard Causes

Much of the contact injury which occurs with robots is due to improper design considerations, while others occur due to improper operational procedures.

Improper design may include controls which are ergonomically unsound, a need for the operator to be within hazardous arcs in order to "teach" the robot, adjust it, or maintain it, all while the robot is able to perform work. There may be no provisions for persons to know or recognize robot/program states. Robots may be tremendously capable of overpower or overspeed for certain task requirements. There may be no provisions for stopping equipment when out-of-performance occurs. Out-of-sequence operations may not be accommodated. Equipment may not be failsafe in design, especially regarding a power glitch or resetting of the program. Structural strength design may be inadequate for static and dynamic loads.

Proper robotics procedures may lack specification, training, or adherence cross-checks. Improper teaching/programming of the robot may occur with unexpected results. Enhancements may be made to the system, without revisiting the safety analyses. Personnel changes may occur, resulting in loss of safety experience.

Control of Robotic Hazards

System-safety applications can improve a robotics safety program. A most common primary control for robotic hazards is that of establishing a safety zone around the robot. Depending upon the energy, speed, and versatility of the robot, this zone may need to be physically protected. Access to this zone is often protected with interlocks to power down or inactivate the robotics if the space is violated. Where multiple robots are employed, zones must not overlap. Zones may be protected by locked rooms, wire cages, interlocked doors or gates, sensed spaces (light barriers, sonar devices, pressure mats, and other sensors).

Another primary control is that of designing to strength and performance-requirement standards. Like any mechanical device, especially one involving mechanical handling aspects, safety factors must be designed in and tests performed to verify that safe performance is present. For instance, robots should be proof tested above their rated loads. Load tests should use the actual end-effector. Speed limitations, such as during teaching mode operation, must be verified. Failsafe provisions must be tested as working properly, using induced faults.

Another primary control is the performance of software safety analysis. Software safety is addressed earlier in this Safety Skills section.

A secondary control, but very important, is that of operator training. Many controls are essentially procedural, so operators must be trained to have the required information and to respond correctly in a variety of conditions. Safety engineers should analyze safety training, to assure that key factors are addressed, and that unreasonable risk is not introduced in the training. Training, like any other organizational function, needs to be monitored as meeting organizational standards of excellence. Training needs "organizational preventative maintenance" in order to operate correctly.

Another secondary control would be that of tactile sensors which stop an operation when contact is sensed. The problem with that is damage may have already occurred from the impact, pinching, shearing, heat, or electrical hazards normally associated with many robotic functions.

SAFETY TRAINING

The American Society for Training and Development reports that ninety percent of U.S. workers get no formal training from employers. Another disconcerting finding is that more than fifty percent of production workers are still trained by on-the-job training (OJT), the most ineffective way of training (La Bar, 1991). Since a high proportion of workers injured or killed on the job are relatively new to the job, training has the potential to impact this large risk.

Safety training is mandated by over 100 OSHA standards (La Bar, 1991) and 40 hours of basic safety and health training is required by MSHA within the first 60 days of employment and 16 hours of annual refresher training (Carr, 1991).

Safety training is not, however, a panacea that will overcome poor design, but training is a vital part of implementing most programs. For training to be meaningful, new knowledge and skills have to be involved. An ERDA report stated that training, in the MORT philosophy of systemic upgrade and repair, is an effective solution to a behavioral problem only if skill deficiencies are involved and people have not and are not performing correctly. Often, behavioral change is insensitive to training (Nertney & Buys, 1976). Benner suggests that flawed training from misdirected safety objectives actually raises risks, instead of reducing them. Benner also strongly believes that use of accident investigations to establish training criteria is improper. He suggests that training goals to influence better outcomes through intervention in loss sequences is far better than trying to "prevent accidents" (Benner, 1990). Merely going through the motions of canned presentations to satisfy a requirement, bores the participants and wastes organizational resources. Proper design of training information and interesting presentations can justify the large expenses of worker time and company training resources. Many regulations are mandating training as part of comprehensive standards to improve workplace safety.

The time-honored tradition of tailgate safety talks and regular safety meetings is giving way to tailored safety training sessions which meet specific training objectives and fit better into today's participative management styles and an emphasis upon total quality of work.

Trainers are not born with innate training skills; they must be prepared to teach. Safety training, for this book, is the education of workers in information and skill prerequisites necessary for the safe performance of tasks. Training, to be effective, needs practical application and evaluation to be sure that the knowledge has been transferred and that work performance is favorably influenced.

The first step in the training process is the establishment of training goals, followed by development of curricula, media, materials, and the preparation of the trainers. The actual presentations involve a small proportion of the time to prepare. There are many tricks to the trade of training presentations. Many of these relate to physical arrangements, keeping the presentation interesting, and use of multimedia approaches to information transfer. The latter steps in training involve documentation, evaluation of training effectiveness, process improvement of previous training, and sampling to determine need for follow-up training. Training is a dynamic process, and if allowed to become static, it loses its effectiveness.

Dale and Nyland's Cone of Learning (Figure 4-6) is a graphic illustration of the increasing value of different learning techniques (Everett, 1989).

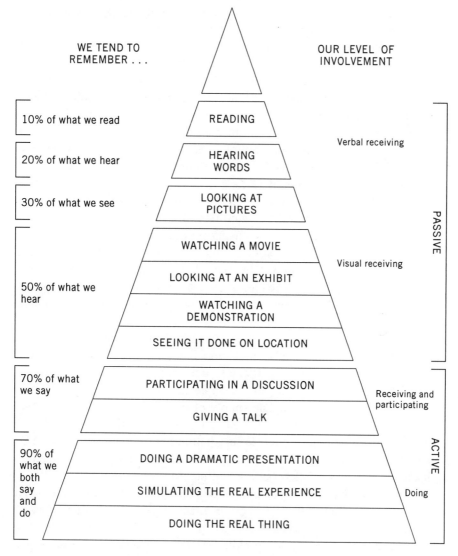

FIGURE 4-6 Cone of Learning. *Source:* Edgar Dale and Bruce Nyland, University of Wisconsin, Eau Claire.

Videotape is a powerful training medium that can be done in-house or by outside professionals. The technology is fairly simple, but professional results are difficult to achieve without a lot of experience. Obvious benefits include instant replay to verify what went on tape, using familiar workplace scenes, and having actual workers as subjects. The difficulty is in scripting, lighting, editing, and title production. Software exists to produce electronic titles and effects. Custom commercial video productions cost around $1,000 per run-minute (Burks, 1990). Many companies rent or sell packaged video productions on a variety of safety and health topics.

Safety meeting presentations lack impact if they are not relevant, interesting, realistic, and applicable (Pater, 1988). Presenters should have training in effective presentation skills, the subject must be carefully prepared, audio/visual aids should be utilized, audience characteristics should be understood, agenda announcements should be issued, audience comfort and receptability should be enhanced, and meetings should be kept brief. Three or four main topics are probably optimal. Scheduling a film or video immediately after a full lunch is doomed for failure to communicate.

The Keep It Simple Stupid (KISS) principle is good advice for most training. "Tell 'em what you're to tell 'em," "then tell 'em," "then tell 'em what you told 'em is helpful," if modified slightly in each repetition. For better presentations, Pater suggests using props, small progressive information blocks; breaks, audience participation, and carefully selected, appropriate humor; loud–soft, fast–slow deliveries; changing distance to the audience, the art of hand gestures and eye contact; use of anecdotes, short testimonials, or linkages to past and future messages, and plans to deal with problem participants or groups (Pater, 1988). OSHA issued voluntary training guidelines (Training Requirements in OSHA Standards and Training Guidelines, OSHA Publication 2254, 1985) which model the design, conduct, evaluation, and revision of occupational health and safety training programs (Samways, 1987). The fourteen Educational Resource Centers which are funded by NIOSH help develop an adequate supply of qualified, properly trained personnel to carry out the OSH Act. ERC programs are built around graduate studies in three of the four engineering, public health, nursing, and medical school core disciplines. NIOSH also funds grants to students from associate's to bachelor's to master's to doctoral and post-doctoral studies at individual schools. The OSHA Training Institute has the purpose of training compliance officers in safety and industrial hygiene, focusing on OSHA regulations.

Training to effect behavioral change, in the MORT concept of system upgrade and repair, must involve skill deficiencies, performance inadequacies must have existed and still exist (Nertney & Buys, 1976). Design of training and procedures themselves often nullify the value of safety or even raise risks instead of lowering them (Benner, 1990). The Nuclear Regulatory Commission's Three Mile Island Special Inquiry Group identified these problems with procedures:

1. Lack of consistency between nomenclature used in procedures versus that on panel components.
2. Instructions for control actions seldom provided indications of correct/incorrect system response.
3. Procedures overloaded operator short-term memory.
4. Chart/graph information needed were not integrated into text.
5. It wasn't always clear which procedures pertained to what situations.
6. There was no formal way to get operator inputs into procedure updates.
7. Procedures were deficient in helping operators in diagnosing events.
8. Procedures were not oriented toward restoration of lost/degraded safety functions.
9. Training on procedures was not closely associated with the procedures (Ostrom et al., 1991).

Ostrom et al. (1991) suggest a dual-column procedure format where each instruction includes an indication (if applicable) and actions to take in case that the response is not obtained. If/then conditional instructions are given, along with preceding warning and caution headers for hazardous steps (Ostrom et al., 1991).

Della-Giustina and Deay (1991) suggest trainer characteristics involving trainer adaptive instruction styles, trainee participation, and trainer behavior overcoming authoritarian style. They also emphasize the high degree of diagnostic power of certain preplanned question forms. Another powerful training development is that of nonlinear access computer simulation which uses videodisc technology, for greater realism (Della-Giustina & Deay, 1991). Branching programs allow the student to tailor the training along individual interest lines.

SAFETY TRENDS

Trends, that is, directional changes, are of extreme importance to proper allocation of limited resources for safety and health. Predicting trends is more difficult than recognizing them. Databases and expertise in analyzing the significance of changes is necessary.

Safety statistics are discussed elsewhere in this text.

It is important to recognize that trends assume more importance if they are temporally related. Changes in raw numbers suggest a trend, but as those playing the stock market know, do not assure projected future performance. Models of system behavior are also needed. Trend prediction in areas where little is known of system behavior, require a great deal of luck or a great deal of insight.

Dugald Pinyan (1988) in an insightful article describing emerging trends, reminds us that if we do not learn from the mistakes of the past, we are doomed forever to repeat them. He related the NSC's Research Projects Committee 1985 study on emerging trends for use by the Industrial Division. A large Delphi Method (expert panel) ranking of about 150 topics produced seven trends of "high" importance:

1. Changing composition/nature of workforce/jobs
2. Occupational vs. non-occupational illness
3. Development of safety standards
4. Defining acceptable risks
5. Improved failure prediction and preventative maintenance
6. Long-term effects of toxics in the environment
7. Defensive medicine

In terms of public interest, only two items rated "very high":

1. Substance abuse
2. AIDS

Five short-term trends were revealed (in descending order):

1. Hazardous waste management
2. Substance abuse
3. Right-to-know
4. Vehicle/air collision avoidance
5. Stress (on or off job)

Finally, seven long-term trends were revealed (in descending order):

1. Changing composition and nature of workforce/jobs
2. Defining acceptable risks

3. The safety/health–legal interface
4. Long-term effects of toxics in the environment
5. Development of safety standards
6. Occupational vs. non-occupational illness
7. Health risk factors

NIOSH compiled this list of ten leading work-related diseases and injuries in the United States (*Occupational Health & Safety,* June 1985):

1. Occupational lung diseases
2. Musculoskeletal injuries
3. Occupational cancers (other than lung)
4. Severe occupational traumatic injuries
5. Cardiovascular diseases
6. Diseases of reproduction
7. Neurotoxic disorders
8. Noise-induced loss of hearing
9. Dermatologic conditions
10. Psychological disorders

Political agendas do not always follow expert opinions. A 1990 BNA report suggested that the following health and safety issues would be important to OSHA, industry and organized labor (*Professional Safety,* Feb. 1990):

1. Ergonomics (especially repetitive-motion problems)
2. Motor vehicle safety (work requirements for seat belts and/or restraints)
3. Indoor air pollution
4. Construction safety
5. Reform of the act
6. Right-to-know: Material safety data sheets
7. Medical surveillance standards
8. Bloodborne diseases (AIDS, hepatitis B, and other infectious bloodborne diseases) and
9. Asbestos (controls, short-duration operations, employees smoking controls, operation-specific limits and bilingual warnings and labels).

One of the projected emerging trends (cumulative trauma disorders) have been observed in the meatpacking industry to be 75 times the rate of industry as a whole (Sheridan, 1991). Workplace illnesses associated with repetitive trauma (repeated motion, pressure, or vibration) were over 50% of surveyed occupational illnesses in 1989 (*AIHA Journal*, Feb. 1991).

Another BNA Special Report addressed "Occupational Safety & Health: 7 Critical Issues for the 1990s," as follows (Bureau of National Affairs, July 1989):

1. Criminal prosecutions
2. Worker notification programs
3. AIDS and other blood-borne diseases
4. Chemical exposure standards
5. OSHA recordkeeping requirements
6. Right-to-ACT initiatives
7. Construction safety regulations

In the author's opinion, there are several additional safety and health trends of concern in the future including:

1. Increased toxic tort lawsuits
2. Increased use of system safety
3. Software safety issues
4. Catastrophe avoidance
5. Restricted workplace smoking

Simple total count safety statistics continue to be mentioned, such as the NSC's 1989 total number of workers killed on the job (10,400), and that there were eight straight years of decrease in this statistic through 1989. The recent 1989 one-year number-of-deaths reduction was about 2% from the previous year and the disabling injury number dropped almost 6% to 1.7 million. Ranking of type of fatal accidents indicated that motor vehicle accidents continue to rank as the most frequent cause (*Occupational Hazards*, Dec. 1990).

Companies are tending to pay attention not only to the numbers of occurrence, but to associated costs, and not just the direct costs that are paid in premiums and payments. Total costs may be as high as 90 times injury costs (Allison, 1991). Detailed information on safety costs are contained in another section of this text. Costs are a surrogate that are trendable.

NASA has initiated formal trend analysis programs such as PRACA (Problem Reporting and Corrective Action). This is to produce vital information for management in two areas: gross numbers of problems and analysis for effects upon systems, subsystems, down to components. Johnson Space Center has further implemented PRACA by the RPTARS (Reliability/Problem Trend Analysis and Reporting System) which is to identify and elevate adverse reliability/problem trends to appropriate program management; to implement comprehensive information management; ensure critical equipment/systems retain design reliability throughout program life; ensure analyzed data application to time/age/life-cycle program requirements, and implementation of closed-loop trending tracking (*NASA JSC*, April 14, 1989). The importance in this example of a trending program is valuable for how it must be applied to the system to achieve change.

Another trend of importance to safety engineering is that of staffing and budgets. For instance, a 1977 example of this was that cost and size of medical/safety staffs had doubled since the OSH Act's institution in 1970. The overall prediction was that size and cost would double in five more years and 44% of participants believed that costs would triple over five years (*Occupational Health & Safety*, Mar./Apr. 1978).

SOFTWARE SAFETY

Software safety is a subset (specialization) within system safety. Software and its interfaces are increasingly important causes of system hazards. Software safety is relatively recent in its development. The use of digital and analog computer technology to perform human or even superhuman functions is rapidly expanding into everyday aspects of life. When properly done, computer systems can approach or exceed human capabilities, depending upon the task. Digital computers are increasingly being used to monitor and/or control complex, time-critical hazardous processes or devices (Leveson, 1986). The use of computers in the space shuttle and in fly-by-wire aircraft are good examples. And, without these computers, the risks escalate rapidly. Thus the computer is used for risk avoidance when it is properly programmed and is properly functioning, yet it can increase serious injury or death occurrence if not.

Nevertheless, this technology is not without safety considerations. The potential for system harm comes about for reasons similar to strictly human-based systems. Like human operators, both errors and "sneaks" (inadvertent design functional errors) can occur due to the use of computers and their related software programs. Fatalities and serious injuries, loss of costly systems, and loss of functional capability are becoming commonplace.

Risk assessment for software control of critical functions is made difficult by the complexity of code, multiple branching, programmed internal decision logic, parallel processing, and computer voting, lack of statusing feedback during operations, and the overall lightning speed of operations. Leveson points out that "Software can have various unexpected and undesired effects when used in complex systems. The software is correct or incorrect, only with respect to some larger system in which it is functioning." Possible hazardous states in a software system may be exceeding large in number (Leveson, 1986).

The hazard cause can be software related, but the software itself is only a contributing cause. The result of improper outputs from a software program or its computer hardware may allow energy exchange or some temporal interaction which a proper sequencing or timing would not produce.

Often, because of the intricacy or the short times involved, humans are unable to detect computer errors or their results, in time to correct. If systems do not have backup computer checking, or do not display human-comprehensible signals, the human in the system is often unable to properly overcome computer-generated actions. Mark Pliakos states that implementing a software function does not allow the normal hardware parameter of rate of occurrence (Pliakos, 1992). Instead, the Schulmeyer Software Safety Risk classification system uses software autonomy and ability of operator intervention prior to damage, in the matrix shown below in Figure 4-7.

Pliakos states that short of reducing software control scope, there is no practical way to move "right and down" in the matrix shown in Figure 4-6.

HAZARD SEVERITY

	I	II	III	IV
Autonomous Time Critical	1 AT	2 AT	3 AT	4 AT
Autonomous Not Time Critical	1 AN	2 AN	3 AN	4 AN
Information Time Critical	1 IT	2 IT	3 IT	4 IT
Operator Control	1 OC	2 OC	3 OC	4 OC
Information Decision	1 ID	2 ID	3 ID	4 ID

FIGURE 4-7   Example of Software Risk Matrix

Also, he stresses the need for good software engineering design such that programs work according to Ian Sommerville's "(software) failsafe criteria." That is, a program should not produce incorrect output; regardless of the input, it should not be corruptable, and it should respond in defined, meaningful, and useful actions when unexpected situations occur. Complete program failure should occur only when further progress is impossible (Pliakos, 1992).

Much of the time, the problems occur at subsystem interfaces, especially those interfaces with humans. Humans are notorious for mistakes in the long run. Other times, the problems are due to improper interfaces with other machines, even other computers. Small discrepancies in the internal clocks of different computers in a system can cause undesired interactive results. Another kind of problem is the introduction of a spurious signal. This generally is a physical problem relating to electrical charge, often at the molecular level. Computer systems are heat and vibration sensitive, as well as dust and radiation sensitive.

Another computer-related safety concern is that of common cause. The more things controlled by a computer system, the more things that can be adversely affected by system failures.

With the advent of artificial intelligence, computers will have more opportunity to make their own thinking mistakes. Programming errors will be harder to detect and it will be even harder to program self-correction.

The solution to hazard control differs little from the logic used for human operators. Functions should fail in the safest mode. There should be ways to detect hazardous situations (safety-critical software involvement). And there should be ways for corrective action to correct inevitable output errors. Where practical, redundancy and inhibiting functions should protect against single-point failures and particularly critical functions.

The analyses for software safety are best done in the design stage. Proof testing should take place using as realistic conditions as is possible. Good software testing is an art, and a difficult art at that. Since software analysis and testing is expensive, software safety should be limited to safety-critical software systems.

Leveson (1986) defines software safety as the assurance that software will execute within a system context without resulting in unacceptable risk. She says that "Inadequate design foresight and specification errors are the greatest cause of software safety problems." Therefore, a system performance viewpoint and front-end analysis are necessary to software safety.

Fault Tree Analysis, Real Time Logic, and Time Petri Nets are examples of technology which have been applied to software safety. There are specific models, such as the "Safe Product Analyses Model" (Askren & Howard, 1988), which attempt to determine software hazard cause and potential hazard controls.

Software safety hazard controls include the following:

1. Elimination of erroneous requirements and development of software requirements which control potential system hazards
2. Designing-out unnecessary human interaction, yet designing-in human control prior to execution of critical commands
3. Designing fault-tolerant software, especially use of multiple inhibits and program checking regarding operator inputs
4. Designing-in optimal failsafe software modes, including error condition and treatment during execution of safety-critical programs
5. Validation of programs to determine software faults and sneaks
6. Review of human factors for machine–operator compatibility, especially in command and control functions
7. Design of operating procedures to incorporate safe interruption and recovery operations
8. Design compatibility with both software and hardware environments
9. Design incorporation of status cues for operator indications of potential machine activation/operation/movement
10. Review of all software interfaces
11. Design human presence detectors capable of protecting against software-controlled machine energies
12. Design guards to prevent contact with uncontrolled machine energies
13. Design program partitions and separation such that crossover between multiple operating modes will not occur
14. Provide warnings of software-related hazards for the uninitiated
15. Provide training for hazard avoidance in software-controlled systems
16. Require formal validation/demonstration of requirements to license safety-critical control software
17. Develop safety standards for software control of critical functions
18. Develop catastrophic system potential for software-related failures
19. Control combinations of relatively common faults' ability to cause hazardous effects
20. Preclude incorrect operator indications from software-driven programs which could cause serious adverse effects

STRUCTURAL INTEGRITY LOSS

Many hazards relate to loss of structural integrity. For this book, structural integrity is that quality that, if failed, may cause catastrophic energy-related effects. Examples include bridge truss member failures, pressure vessel wall failures, high-pressure line or connection failures, rocket motor support failures, and scaffold structural member failures. Failures can involve physical breaking, rupturing, or detrimental deformation.

Care should be exercised about what the ultimate design loads should be during all phases of system life for combinations of pressure, thermal, and mechanical loads. Low-frequency and random vibration loads should be considered, especially with regard to critical frequencies. Both static and dynamic loads analysis must be performed to assure that maximum anticipated loads can be safely tolerated.

In general, load-bearing structural members can fail due to insufficient strength and/or stiffness, improper manufacturing or assembly, inadequate or flawed materials, damage to structures, improper loading (overload, compressive instead of tensile forces, excessive cycling, pressure, and thermal loads, etc.) and fracture failures such as stress corrosion cracking (prolonged tensile stress with corrosion), incompatible material contact (galvanic action), stress concentrations, and use of counterfeit or improperly secured fasteners.

NASA uses "fracture control" systems engineering approach to control catastrophic loss of structural integrity. Undetectable flow growth is controlled over the entire system life of structural articles by design of safety margins and factors of safety over limit loads. Factors of safety are used to compensate for uncertainties in material properties and in analyses regarding strength and strain. Probability distributions of loads may be necessary where time-consistency of loads is unknown.

Good engineering design practices are required in fracture control, such as minimizing eccentricities and stress concentrations to avoid initiating fatigue crack propagation; providing proper access to perform assembly, test and maintenance activities; selecting proper materials for enviroments and use; use of adequate nondestructive analysis techniques to identify unacceptable flaws; addressing damaging effects such as hydrogen embrittlement, temper embrittlement, creep, general and galvanic corrosion, eutectic melting, radiational damage, and atomic oxygen effects; and providing contained or restrained or failsafe designs, where possible, to achieve failure tolerant design. NASA's MIL-HBK-505 and MSFC JA-418 contain excellent guidance on fracture control.

Stress corrosion control can be accomplished by careful selection and control of materials, weldment techniques and fillers, proper heat treating, and design for cumulative stress throughout all phases of manufacturing and assembly, transportation, testing, storage, and use. For example, austenitic steels of the 300 series are less stress corrosion-susceptible than the martensitic steels of the 400 series. But with both series, heat treatment must be closely considered. High nickel alloys tend to be less susceptible. Higher strength aluminum alloys are often susceptible and should be mechanically stress relieved. Surface treatments must be analyzed for their potential adverse effects. NASA's specification MSFC-522A has excellent guidelines for stress corrosion control. Also, length of service in particular environments strongly affects stress corrosion cracking. Both transverse and longitudinal stress must be considered in designs regarding stress cracking.

Verification of safe designs requires analysis, testing, and demonstration that systems can meet system requirements. Verification examples include Safe Life Testing, Proof Testing, Static Testing, Pressure Testing, Failure Testing, and Modal Vibration Analysis.

Fasteners introduce additional points for structural integrity to be lost. The fasteners themselves may fail similarly to main components of the structure, in ways previously described. Also, fasteners may be properly designed, but not be made according to requirements, such as "counterfeit" fasteners substituted illegally. Further, good fasteners of varying capabilities may be inadvertently mixed due to poor drawings or assembly procedures. Good fasteners may be damaged during installation and/or removal. Torque must be carefully controlled and preloads must be limited. Formal counterfeit fastener control programs have been established to avoid inadvertent use of inadequate fasteners of this type.

In closing, engineering of structural integrity is indeed a mechanical engineering discipline in itself. Further details are beyond the scope of this book. Safety engineers may be expected to use various levels of expertise in this skill area. There should be close coordination with highly qualified experts where safety engineers lack structural engineering expertise.

SYSTEM SAFETY

Background

System Safety doesn't have to be difficult, especially if used as a qualitative tool. System safety (alternately known as systems safety) is system engineering applied to safety. It is a convenient logical base for optimization of the safety process for systems, especially complicated and interactive systems. System Safety has the great advantage of assessing safety without the need to analyze loss experience retrospectively. System safety is best used as a forward-looking technique, because after-the-fact change is more expensive.

System safety had its real beginnings in electronics and in weapon systems in the late 1950s and early 1960s. These systems were very complex and very costly. System safety concentrated effectiveness through a systems approach and better assured an acceptable level of safety for the system. System safety technology was developed to review systems consistently in a generic manner. Systems seem to be getting more complicated and more costly, rather than the other way around, so there is a fertile field for system safety application.

Safety vs. Reliability

An important distinction must be made between safety and reliability. Reliability is the assurance that items can perform required function(s), for a specified time, under specified conditions. Safety is the acceptable freedom from injury, damage, loss of resources, or loss of system availability. Note that the degradation or losses are usually limited to catastrophic or critical effects. Reliability and safety are different disciplines which can improve system performance. But, safety has the distinction of being concerned with unacceptable, adverse effects, while reliability is concerned with individual functions, however specified. Many systems can be reliably unsafe because of misspecification, unrecognized interactions, or operator control errors. Reliability presumes that function inputs are proper and present, while safety presumes that worst-case conditions will occur. Hazard controls are designed to avoid or minimize hazardous effects, thus minimizing risks of hazards. Hazard controls use design specifications, redesign, safety margins, different method (unlike) redundancy, simple redundancy, inhibiting of hazardous energy release, energy containment, safety devices, procedural controls, personal protective equipment, and warning systems. Safety and reliability *are* different and may not be equated.

Basic Approaches

System safety analytical methods take two basic approaches: the inductive matrix approach and the deductive tree approach. The intuitive logic that people use instinctively is a combination of the two methods (inductive and deductive reasoning).

The power of the inductive approach to leap from one cause to many effects is complemented by the interrelated deductive chains which lead to the top undesired event.

INDUCTIVE REASONING

- Starts with parts or elements
- Asks, if this element or part failed, what would happen?
- Proceeds from "one" to "many" (particular to general case)

The deductive approach plots backward from the undesired, following each and every path through the possible enabling events. As with any analysis, decisions must be made about where to cut off the further investigation as trivial or unaffordable.

DEDUCTIVE REASONING

- Starts with the Undesired Event
- Asks, "How could this happen?"
- Proceeds from "many" to "one" (general to particular case)

The "Systems Approach" has long grappled with how to optimize outputs by manipulating subsystem variables. This implies that system performance is known, otherwise the effects of manipulations would not be able to be compared with the previous state. As hinted earlier, bounding the system is one tricky aspect of being able to work a system. If bounded too tightly, the solution may be constrained too tightly to consider a good answer to the problem. If bounded too loosely, every answer seems relatively acceptable. Definition of the boundary of a system is, by necessity, arbitrary to some degree (Westchurchman, 1968). Importantly, the environment is "beyond the systems control," yet it affects systems outcomes and therefore must be considered (Westchurchman, 1968).

Once the system is bounded (often a critical step which is very difficult to optimize), and system performance is known, the subsystems and their interrelationships must be understood by the analyst. Components of a system need to be thought of in functional or mission-oriented terms (Westchurchman, 1968). As well, a set of requirements or expectations must be used to guide the analyst. System descriptions, flows, and organizations are needed by the analyst.

Another important aspect is contained in the following description: A system approach is methodical, objective, probably quantitative (at least measurable) and analytical; shows interaction of all constituent elements at any given time, considers all constituent elements as subsystems in parallel, rather than series; describes both inputs and outputs in clear language, and views the entire system as a closed loop. System safety seeks to optimize system performance with respect to safety, while at the same time, being constrained by operational effectiveness, time, and cost. System safety applies engineering and management techniques throughout all phases of a project (from conception to disposal).

System Safety Plans

System safety begins with the development of a System Safety Program Plan or a System Safety Plan. The plan covers scope, resources, program milestone activities and schedules, requirements to be followed (including required analyses and documentation, and updating requirements), and qualifications required for system safety analysts/engineers. Normal management planning formats are acceptable for inclusion of system-safety program-planning information.

Matrix Analyses

The earliest safety analysis is usually a conceptual preliminary hazard analysis (CPHA). A general knowledge of the system conceptual design is sufficient to perform a simple matrix hazard analysis of the "if this thing happens, what could result?" type. System specifications, diagrams, and descriptions should be reviewed and if possible, the design engineers should be consulted about system design intents. Checklists are often used systematically to help identify hazards and hazardous operation relationships to include in the matrix. An example PHA (conceptual or other) matrix format is given using a hot water heater example (see Figure 4-8).

The matrix is rows and columns. The rows contain different items or function hazards, and the columns contain progressive information about hazardous events, causes, effects and severities, probability of occurrence, controls, verifications of controls, and remarks. The column headings prompt entries in each row. There is a logical progression from left to right in the matrix. Row information can be selected from system components and hazard checklists, or analyst experience. Guidance from NASA 22254 describes the column information (see Figure 4-9).

Two of the matrix headings require further explanation. The first, Severity of the hazardous effect (often called hazard level or criticality category) is categorized in the following example (see Figure 4-10).

Hot Water Htr PHA Part. Example

Item/ Function	Hazardous Condition	Cause (s)	Effects	Sev.	Prob.	Recomm./ Verifications
1. Natural Gas Supply	1. Leak	1.1 Valve Faults	Fire or Explosion	Cat	E10-3	Qualified Valve Design, Installation, Pressure Testing, Inspection
		1.2 Line Integrity Faults	Fire or Explosion	Cat	E10-4	Qualified Line/ Connector Design, Installation, Pressure Testing, Inspection
		1.3 High Supply Pressure	Fire or Explosion	Cat	Remote E10-5	Pressure Relief
2. Pilot Flame	2.1 Flame loss allows unburned gas	2.1.1 Supply Low Pressure or Loss	Secondary Fire or Explosion	Cat	E10-3	Low Pressure Cutoff; Control logic for pilot and main gas flow
		2.1.2 Flame sensor fails to cut off gas	Secondary Fire or Explosion	Cat	E10-6	Secondary temp. sensing cutoff of main gas
	2.2 Intermittent Flame	2.2 Intermittent Contamination of gas supply	Secondary Fire or Explosion	Cat	E10-5	Control logic for flame sensing and main gas flow
		Gas Supply Pressure Surges	Secondary Fire or Explosion	Cat	E10-4	High reliability pressure control; Supply pressure cutoff logic
3. Electrical Controls	3.1 110v. ac ground current	3.1.1 Fault to ground	Electrocution Death	Cat	E10-4	Qualified design, Installation, Testing, Inspection
		3.1.2 Improper maintenance	Electrocution Death	Cat	E10-3	Qualified procedures, maintenance personnel
		3.1.3 Control Fault	Electrocution Death	Cat	E10-5	Qualified design, Installation, Testing, Inspection
etc., etc., etc.						

FIGURE 4-8   An Example PHA Matrix Format Using a Hot Water Heater

The second heading is Probability of Occurrence of the hazardous effect category, as shown in Figure 4-11.

It should be recognized that both Severity and Probability can be defined into different categories with different values. The process is what is important, not the absolute values which are used as the basis of a category.

PHA No.: _____ i.e. Ground Ops; Flt Ops

Mission Phase: _____

Analyst: _____

Date: _____

Organization: _____

Sheet No.: _____

System/Subsystem or Operation: i.e. Electrical; Propulsion

Effectivity: i.e. On-orbit; Landing

HAZARDOUS CONDITIONS	HAZARD CAUSES	HAZARD EFFECTS	SEVERITY LEVEL	SAFETY REQ'TS	HAZARD CONTROL	VERIFICATIONS	LIKELIHOOD
1. Hazard(s) as designed? 2. Effects of failures? 3. Timing hazards? 4. Human errors? 5. Generic hazards? Use other checklists? Crosscheck other analysis as inputs?	1. What creates hazard? 2. What conditions? 3. Check using FTA? 4. Hardware and/or Software Use checklists? Crosscheck other analysis as inputs?	1. Related effects of each cause 2. Effects on people and equpt. 3. Effects before controls are applied 4. Include Worst Case Use Checklists?	1. Level for each cause 2. Severity before control CA; CR; MA For each cause	1. List Req'ts that eliminate or control condition Doc. No. and paragraph	1. Means to control hazard cause 2. Use hazard reduction precedence order Keyed to each cause	1. Methods to verify each control 2. Reference documentation (Doc. Number, Title and paragraph) Analysis; tests; procedural requirements	1. Probability of occurrence 2. Without and with hazard controls in place Prob.; Infreq.; Remote; Improb.

FIGURE 4-9 Preliminary Hazard Analysis Instructions. Adapted from Space Shuttle PHA Instructions. *Source:* NASA NSTS 22254, Methodology for Conduct of NSTS Hazard Analysis.

Description	Category	Mishap Definition
Catastrophic	I	Any condition which may cause a permanent disabling or fatal personnel injury, or loss of one of the following: the launch or servicing vehicle; manned base; any NSTS cargo element, the loss of which could result in the loss of the manned base; major ground facility or critical support equipment.
Critical	II	Any condition which may cause a serious personnel injury; severe occupational illness; loss of safety monitoring, emergency control function or an emergency system, or requires use of emergency procedures; or involves major damage to one of the following: the launch or servicing vehicle; manned base; any NSTS cargo element, which could result in the loss of, or major damage to, a major SSF element; an on–orbit life–sustaining function; a ground facility; or any critical support equipment.
Marginal	III	Any condition which may cause major damage to a safety monitoring, emergency control function or an emergency system, mishap of a minor nature inflicting first aid injury to personnel, or minor damage to one of the following: a launch or servicing vehicle; the manned base; any NSTS cargo element, which could result in minor damage to a major SSF element; an on–orbit life–sustaining function; a ground facility; or any critical support equipment.

FIGURE 4-10   Severity of the Hazardous Effect

DESCRIPTION	CATEGORY	MISHAP DEFINITION
PROBABLE	A	EXPECTED TO HAPPEN IN THE LIFE OF THE PROGRAM
INFREQUENT	B	COULD HAPPEN IN THE LIFE OF THE PROGRAM. CONTROLS HAVE SIGNIFICANT LIMITATIONS OR UNCERTAINTIES
REMOTE	C	COULD HAPPEN IN THE LIFE OF THE PROGRAM, BUT NOT EXPECTED. CONTROLS HAVE MINOR LIMITATIONS OR UNCERTAINTIES
IMPROBABLE	D	EXTREMELY REMOTE POSSIBILITY THAT IT WILL HAPPEN IN THE LIFE OF THE PROGRAM. STRONG CONTROLS ARE IN PLACE

FIGURE 4-11   Probability of Occurrence of the Hazardous Effect

Risk Ratings

Severity and Probability factors are used to represent risk ratings. Two examples of matrices combining severity and probability are given in Figure 4-12.

Risk ratings, such as "risk index categories" or "risk assessment categories," are used to guide management action.

Logic Tree Analysis

If particular hazards of interest are identified, a simple top-level fault tree can be developed from an undesired top event. A simple example for "death from hot water heater" is shown in Figure 4-13.

First Example Risk Matrix
RISK ASSESSMENT CODES

	I	II	III	IV
A	1A	2A	3A	4A
B	1B	2B	3B	4B
C	1C	2C	3C	4C
D	1D	2D	3D	4D
E	1E	2E	3E	4E

Second Example Risk Matrix

	I	II	III	IV
A	1	3	7	13
B	2	5	9	16
C	4	6	11	18
D	8	10	14	19
E	12	15	17	20

Hazard Risk Indices:

UNACCEPTABLE (Shaded areas):
        RACs 1A-1C; 2A-2C; 1-5
UNDESIRABLE (Management Decision Required):
        RACs 1D; 2C-2D; 3B-3C; 6-9
ACCEPTABLE (with Management Review):
        RACs 1E-2E; 3D-3E; 4A-4B; 10-17
ACCEPTABLE (without Management Review):
        RACs 4C-4E; 18-20

FIGURE 4-12 Examples of Matrices Combining Hazards and Probability. Adapted from MIL-STD 882B.

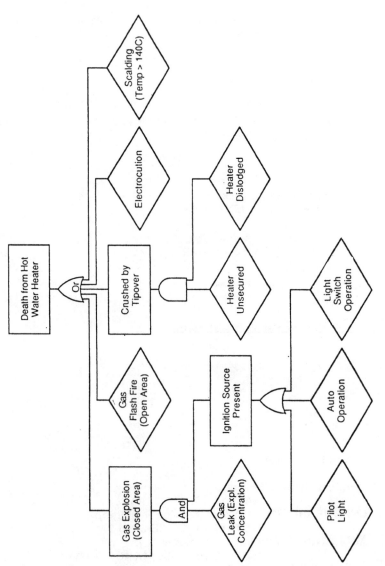

FIGURE 4-13 Example of a Fault Tree

Fault trees are usually drawn with an undesired event at the time and ways that the event could happen beneath, connected to the upper event by logic gates. There are basically only two types of logic gates: "or" and "and". The "or" gate means that any of the contributing events would be sufficient to cause the upper event. The "and" gate means that all of the contributing events must be present for the upper event to occur. There are modified logic gates to take care of certain conditions, such as an event which must be preceded by a certain sequence of events. Further details of fault tree technology is beyond the scope of this text. There are also many more recent texts which address fault tree analysis, some listed in this book's section on Safety References.

Fault trees are most useful when limited to certain events, by careful selection and wording of the top event. Another way to limit the trees is to "diamond off" (terminate the branch using a single diamond symbol) the branches as of no further interest (the causes are well understood and well controlled). Branches where insufficient information is available to continue to lower levels may be marked with a "double diamond" (diamond within a diamond). This reminds fault tree analysts to revisit those portions of the tree at a later date.

Caution must be given to maintaining the correct logic within the tree. Without correct logic, the tree forms incorrect loops of circular logic or merely doesn't portray reality.

Iteration

PHAs can and should be repeated with updated design information, several times during the design process, later involving operational information. This is an example of a common iterative system process of going back and updating/correcting an analysis based upon newer information. This is also true of the need to revisit existing fault trees.

Expanded Matrix Analyses

By changing the matrix headings (asking slightly different questions), similar matrices to the PHA are produced, such as the Subsystem Hazard Analysis (SSHA), the System Hazard Analysis (SHA), and the Operating and Support Hazard Analysis (O&SHA). Example matrix headings are given in Figure 4-14 for three additional matrices.

By adding probabilities of event occurrence to fault trees, a quantitative probability of occurrence of each event may be calculated, usually by a computer program which simplifies the need to perform Boolean algebra.

There are many specialized analysis techniques applicable for system safety, including:

1. Bent Pin Analysis
2. Cable Failure Matrix Analysis
3. Cause–Consequence Analysis
4. Catastrophe Analysis
5. Change Analysis
6. Code-Level Software Hazard Analysis
7. Common Cause Analysis
8. Comparison-To-Criteria Analysis
9. Contingency Analysis
10. Crew Safety Hazard Analysis
11. Critical Incident Technique
12. Critical Path Analysis
13. Criticality Analysis
14. Detailed Design Hazard Analysis
15. Energy Analysis

PHA/SSHA/SHA

ITEM/ FUNCTION	HAZARDOUS CONDITION	CAUSE	EFFECTS	SEV.	PROB.	RECOM./ VERIFCN.

OHA

PROC./TASK OPERATION	HAZARDOUS CONDITION	CAUSE	EFFECTS	SEV.	PROB.	RECOM./ VERIFCN.

FIGURE 4-14   Example Matrix Headings

16. Energy Flow/Barrier Analysis
17. Energy Transfer Analysis
18. Event Tree Analysis
19. Failure Modes and Effects Criticality Analysis
20. Fault Hazard Analysis
21. Fault Isolation Analysis
22. Flow Analysis
23. High Potential Analysis (HiPo)
24. Human Error Analysis
25. Integrated System Hazard Analysis
26. Interface Hazard Analysis
27. Job Safety Analysis
28. Maintenance Hazard Analysis
29. Man-Path Analysis
30. Management Oversight & Risk Tree Analysis (MORT)
31. Multilinear Event Sequence Analysis
32. Naked Man Analysis
33. Network Logic Analysis
34. Nuclear Safety Cross-Check Analysis
35. Operating Hazard Analysis
36. Petri Network Analysis
37. Procedures Analysis
38. Profile Analysis
39. Root Cause Analysis
40. Sneak Analysis and Sneak Circuit Analysis
41. Software Hazard Analysis
42. Software Change Hazard Analysis
43. Software Requirements Hazard Analysis
44. Software/User Interface Analysis
45. Strategy Selection
46. System Hazardous Failure Mode Analysis
47. System Malfunction Effects Analysis
48. Technique for Human Error Prediction (THERP)
49. Test Safety Analysis
50. Time Sequencing Analysis
51. Worst-Case Analysis

Detailed explanation of the preceding partial list of techniques is beyond the scope of this text. Software safety is considered important enough to be treated in a separate section of this book. There are many more techniques not shown here, and other new ones which are created regularly.

Once the analysis is done to identify hazards, qualified and/or quantify the risk, and state hazard controls and verifications, the results need to be recorded in hazard reports and risk-assessment reports. Safety assessment reports are commonly used to highlight the analysis results and give input to the necessary management decisions which are called for by the analysis.

Programs often used phased safety reviews to build in recurrent review of safety at key program milestones. These are typically a Preliminary Safety Review at 10–30% of design completion, a Critical Design Safety Review at about 90% of design completion, and a Design Certification Safety Review at 100% design completion. Programs may also include Operational Readiness Reviews, such as a Flight Readiness Review for aerospace applications. Safety reviews may be specified in contracts and/or in System Safety Program Plans.

The more formal program safety reviews produce a specified safety data package, require formal data package reviews by various disciplines, require standardized discrepancy reporting (such as use of Review Item Discrepancies), and require formalized presentations to review panels and/or boards. Boards or panels issue action items to correct discrepancies, require additional studies, occasionally call for additional reviews (delta reviews), approve acceptable hazard reporting, and approve deviations, waivers, and certificates of safety.

TRANSPORTATION SAFETY

Transportation modes include land, water, air, and space. Mechanized land-based transportation then involves a wide variety of powered motor vehicles including all sorts of automotive and rail vehicles, but less apparent vehicles such as tracked vehicles, cable-driven tramways, traveling straddle cranes, and many other on- and off-road specialized vehicles. This book section shall be limited to automotive vehicle and truck transportation safety.

Transportation involves a great many safety concerns in the movement of people and other objects. Transportation safety is much more than the driver education and fleet vehicle maintenance emphasis that many safety programs provide. Transportation safety is extremely important in the overall, because 39 percent of all occupational fatalities occur in motor vehicles, both on and off the highways (*Occupational Hazards*, Dec. 1989). In fact, motor vehicle crash is the leading on-the-job cause of death. OSHA says that motor vehicle crashes cause 2,100 deaths and 91,000 lost-workdays annually (*Occupational Hazards*, Sept. 1990).

According to National Traffic Safety Administration statistics, the 1989 vehicular death rate was 2.2 deaths per 100 million miles traveled, or about 45,500 deaths per year for the entire annual travel in the United States. The ten-year fatality rate reduction from 3.3 in 1979 to 2.2 in 1989 is considered a marked improvement (*Professional Safety*, Feb. 1991). In almost 34 million total annual automobile accidents, 5.4 million people are injured (*USAA Aide*, Oct. 1989). Traffic accidents annually generate 5,000 spinal cord injuries, 180,000 brain injuries, and 625,000 facial lacerations or trauma (McDermott, 1985). Further, many of the epilepsy cases are caused by these head injuries (Clark, 1984). There has been a tenfold decrease in death rate since statistics were kept in 1921. Part of the decrease is said to be from better safety standards, including highway design, automobile design (such as anti-lock braking), along with seat belt and air bag use. Many believe that raised minimum drinking age laws have helped significantly. In 1988, fully 50% of traffic deaths involved alcohol (*Professional Safety*, Oct. 1989). Also, about 35% of drunk driving accidents involve young drivers from age 16 to 24 (*Traffic Safety*, Nov./Dec. 1982). The National Transportation Safety Board (NTSB) believes that administrative license revocation procedures for driving under the influence of drugs is one of the most effective measures toward reducing drug-related crashes (*Professional Safety*, Oct. 1989). There seems to be little correlation between current driver's education and improved driving performance. All these points draw attention to the significant and the not as significant aspects of transportation risk and its control. Additional factors are contained in the paragraphs below.

Seatbelt use (lap and belt used together) clearly reduce risk of death 40–50%, saving victims from ejection which is involved in almost 25% of car fatalities. Air bag protection alone reduces fatality risk by 20–40%, while seatbelt and air bag combination reduce this risk 45–55% (*Professional Safety*, Oct. 1989). Two hundred million miles of experience and 114 Ford Motor Company air bag collisions resulted in only one fatality, and that one was considered hopeless for survival (Weiss, 1988). Air bag protection is ineffective for side crashes (which represent about 50% of crashes).

About half the states and most federal facilities now mandate seatbelt use. Currently, fifteen times as many front seat occupants die, as those in rear seats. NHTSA is now requiring automakers to provide rear seat lap belts and shoulder belts in new cars, and may extend this to vans, utility vehicles, light trucks and convertibles (*Professional Safety*, Aug. 1989). Unfortunately, nationwide use of seatbelts is only about 50%. Poor adjustment of the seatbelt in an ill-advised quest for "comfort" and in part due to poor design, detracts from seatbelt effectiveness. Seatbelts worn incorrectly under the arm can induce injuries (O'Neill, Apr. 1988).

Most transportation systems have a heavy reliance upon a human operator although increased automation and remote control is evident in recent times. Many of the transportation modes occur in hostile environments and may as well involve high velocities and/or mass-energies.

Twenty-nine percent of work-related crash fatalities involve large trucks, due mainly to their large mass and high undercarriages (Abercrombie, 1983). The federal Motor Carrier Safety Assistance Program (MCASP) provides assistance to states by funding roadside inspections and better enforcement of truck safety regulations. The computer database called Safetynet helps identify problem drivers, vehicles, and carriers. Large trucks, transporters of more than 14 passengers, and hazardous materials transporters are covered by the vehicle inspections under the Tandem Truck Safety Act of 1984. NHTSA may soon require trucks to have controlled collapsing frontal structures and side/rear crash guards to reduce risk to the smaller automobiles. Commercial driver abuses relating to multiple drivers licenses to avoid suspensions, and noncompliance with on-duty hour limitations have been addressed by new laws. Training and certifications are required for hazardous materials haulers. Mandatory drug testing requirements are being imposed upon motor carriers. Federal plans for stiff penalties for convictions of driving under the influence of drugs/alcohol have been proposed (Cross, 1990). About 48% of those jailed driving while intoxicated (DWI) had previous DWI convictions (*Professional Safety*, Aug. 1988).

Visibility contributes to crashes in that about 90% of driving information comes visually. Lighter vehicle colors are seen better, and therefore are less involved in crashes. White is twelve times easier to spot than black, and lighter shades of yellow, orange, blue and grey are safer colors (*USAA Aide*, Oct. 1990). Daylight headlight use may reduce frontal crash rates by up to 22% (Kenel, 1989).

Another significant factor in crashes is that highways are, for the most part, designed for automobiles rather than trucks. Turning radii, passing lane distances, stopping distances, interchange ramps, are designed for passenger cars. Many of the overturning and jackknifing crashes may be related to application of the higher car performance limitations by trucks. The Office of Technology Assessment suggests that design improvements are needed for lane and shoulder widths, roadside and side slopes, bridge entrance widths, road horizontal alignments, sight distances, intersections, pavement edge drops, pavement surface conditions, posted speeds for ramps and curves, and sign design, height, and location (*Professional Safety*, Mar. 1989).

Fatigue is a recognized contributor to vehicular crashes. Insurance Institute for Highway Safety researchers found that drivers who had been behind the wheel eight hours (up to ten hours driving after eight hours rest is legal) were twice as likely to be in a crash that drivers on the road less than two hours (*Professional Safety*, Dec. 1987).

The Federal Highway Administration (FHWA) is proposing minimum testing and licensing standards for all states and the District of Columbia for interstate and intrastate commercial truck and bus drivers (*Professional Safety*, Feb. 1988).

The number of accidents (crashes) per individual over a three-year period is reported to be a reliable predictor of future vehicle mishaps. However, Motor Vehicle Record (MVR) points assigned for operator infractions, is not (Bonner, 1987).

Retrospective relative studies of car safety performance are prepared annually by the Insurance Institute for Highway Safety (IIHS) and the Highway Loss Data Institute (HLDI). A relative injury claim frequency rates cars by make and model where sufficient data are available. Car size and driver's age/sex appear to influence differences in death rates. Only one small car rated among the top ten (*USAA Aide*, Apr. 1990).

Design improvements include better tire designs, vehicle suspensions which lead to better handling, crashworthy energy-absorbing front end collapse, third brake lights, energy absorbing barriers, breakaway signs, and improved directional signs. Improved car inspection programs helps assure a minimal, safely maintained vehicle condition. Anti-lacerative windshields (with a layer of plastic bonded onto the occupant side of the window) should be a great improvement over the highly penetration-resistant windshields which vehicles now have. The existing windshields have helped the ejection problem and the anti-lacerative windshields should help the major injuries that now require extensive plastic surgery.

A problem related to large vehicles, often specialized powered vehicles, is slips and falls during the mounting and demounting from cabs. The access/egress hazards account for about one-fourth of all driver injuries. Ergonomically unsound designs and environmental conditions contribute to this problem. Hidden steps, non-uniform steps, high steps, and single or no handrails are major examples (Hurst & Khalil, 1984).

Being struck by vehicles is a serious problem for pedestrians and for motorcylists. Pedestrian deaths have been estimated to be over 7,000 per year (Texas Traffic Safety Report, 1980). Also, 50% of pedestrian deaths involve intoxication. Motorcyclist death rates are 10–15 times as great as automobile riders (*USAA Aide*, Fall 1986). Motorcyclists crashes are most often caused by cars abruptly turning into their paths in an intersection, due to car drivers not seeing the cyclists. Cyclists using headlights during the day and mandatory use of helmets have reduced crashes, deaths, and serious injuries. Motorcyclist deaths rose 41% from 1975 to 1978 when many states repealed or limited mandatory helmet use (Abercrombie, 1980) and then dropped 30% in Louisiana when the law was reinstated in 1982 (*USAA Aide*, Fall 1986).

Servicing of large vehicles, in particular, servicing of truck tires has a high severity of injuries.

Another vehicular safety-related topic is that of injuries from improper battery jumping resulting in a hydrogen explosion and splattered battery acid.

Safety engineering control of transportation safety hazards and risks include (1) ergonomics man–machine–environment improvements such as control design, ergonomics, emergency actions training, (2) vehicle design improvements such as steering, braking, crash survivability and vehicle visibility aids, (3) roadway design such as visibility considerations, road surface, crown, hill, and curve radii, nominal driver reaction time, spacing of signs, loads, grade crossings, (4) driver training simulation systems to improve driving skills, (5) procedure development for hazardous transportation-related activities such as in vehicle maintenance, and (6) functional operator personal protection such as helmets for motorcyclists.

WALKING AND WORKING SURFACES

Walking and Working Surfaces (W/WS) focuses upon the working surface as contributing directly to the hazard. For example, floors and aisles could be hazardous due to low friction such as water and oil or slick wax. Parking lots could have uncleared snow and ice which contribute to slips and falls. Foreign objects could represent trip hazards where visual cues are not sufficient to allow detection and avoidance. Moving walkways, escalators, man-lifts, and ski lifts could be unforgiving of miscalculations or miscues. Continually moving conveyances such as these have the additional hazard of requiring correct action to dismount, as well as to mount.

W/WS is a safety problem due to the potential severity of injuries due to falls, second only to motor vehicles (and the unclassified "other") as a fatality cause (13% of accidental deaths in 1979). Another consideration is that various W/WS are a continual exposure source during work and some W/WS involve elevations and even Immediately Dangerous to Life/Health (IDLH) conditions. Elevated working surfaces are discussed further in the Construction Safety section. IDLH conditions are discussed further in the Confined Spaces section.

According to statistics, stairs, steps, and ladders are involved in many injuries and deaths. Applicable standards in OSHA, ANSI, and NFPA Life Safety Code 101 would, if followed, eliminate many of these losses.

Another major control besides basic codes above for W/WS would be to control reduced coefficient of friction. This is further discussed under the Fall Protection section. Choice of W/WS surfaces can accentuate coefficient of friction problems. For instance, the common use of diamond plate dramatically reduces the area in contact with footwear. However, there are personal protective equipment footwear sole materials and designs which provide improved traction on slippery surfaces.

It is not practical to individually address all aspects of W/WS safety, due to the wide variety of W/WS, such cab interiors of overhead cranes, building structure during construction, outdoor surfaces, platforms and seats on a variety of vehicles, and underground mine surfaces. The characteristics of W/WS should be reviewed such as with a hazard analysis, covering accessibility, compatibility with the worker, ability to egress, and other factors.

WASTE HANDLING

Waste handling contains many operations with safety and health concerns. For this book, waste handling includes the collection, storage, treatment, disposal, and cleaning up of hazardous waste materials. There is a considerable amount of overlap with information included in the Hazardous Materials section. Not all hazardous materials are wastes.

The Environmental Protection Agency (EPA) estimates that the U.S. generates 570 million tons of hazardous waste annually (R. B. Smith, 1990). Waste materials include toxic, biologic, and radioactive waste. Safety engineering is concerned with human interaction with all aspects of this sanitation activity, considerably broader than concerns with sanitary sewage treatment and sanitary landfill operations. Safety and health concerns may lie within classical environmental spans of control.

Industrial waste includes the following classifications:

1. Abrasives
2. Acids
3. Animal products/by-products
4. Bases (alkalis)
5. Biologic substances
6. Carcinogenic substances

7. Combustible materials

8. Explosives

9. Flammable materials

10. Herbicides

11. Metals

12. Oils

13. Organic materials

14. Pesticides

15. Radioactive substances

16. Reactive materials

17. Salts

18. Solvents

Environmental legislation (Clean Water, Clean Air, etc.) has been enacted to control harmful agents entering the biological cycles that eventually affect humans. Wastes become pollutants when they adversely affect other persons or degrade or damage the ecology. Regulations apply to most of the hazardous wastes which are normally produced by industries. Contents of sanitary sewers and storm sewers must be closely monitored. Amounts generated, on-site storage, transfer to others, and so forth, are reportable items. Of particular concern are highly toxic, persistent, radioactive or reactive materials.

Lists of "hazardous" materials are quite dynamic in their makeup, so reference to Environmental Protection Agency lists is necessary. Chemicals are listed with chemical names and Chemical Abstract Service (CAS) numbers for clarity due to a variety of common names which may overlap or have multiple meanings. From a safety standpoint, not all "hazardous" materials will cause direct harm to all people. But safety engineers should take advantage of these listings to help identify the materials which can cause direct harm to people.

Solid waste disposal uses the following methods:

1. Burning
   (a) Open burning
   (b) Incinerator burning
2. Burial*
   (a) Landfill

*With or without pretreatment, encapsulation, compaction, etc.

(b) At sea

3. Transfer to commercial services for disposal

Liquid waste/sludge disposal uses the following methods*:

1. Deep-well injection
2. Conversion into a solid for disposal
3. Burning/incineration
4. Oxidation, aeration, and air stripping
5. Biological treatment, bioremediation
6. Discharge into water (dilution)

Gaseous waste disposal uses methods such as these:

1. Burning
2. Scrubbing
3. Absorption
4. Dispersal into atmosphere
5. Conversion into less toxic materials (chemical and biological)

Many industrial wastes are sludges with varying amounts of solids in a liquid. Environmental concerns include pH, toxicity, bioenriching agents, heavy metals, and others. Biological treatment microorganisms utilize fermentation, aerobic respiration (redox) reactions with oxygen as an electron acceptor, and anaerobic respiration reactions with sulfates/nitrates/carbon dioxide as an electron acceptor (Novitski & Trattner, 1991).

Filtering, precipitation, and flocculation are used to remove particulates. Microencapsulation of particles by chemical reactions are used. Possolans (aluminous and siliceous compounds) such as in volcanic ash or in industry's power plant fly ash, cement making or iron/steel making by-products, are reacted with lime/lime-like materials to form nearly impermeable masses (similar to cement) that hold waste particles. Both wet and dry forms of the microencapsulated materials can be produced that resist leaching in landfills or road use (Metry & Fazzini, 1980).

Carl Adams (1991) put together the following excellent comprehensive waste treatment technology list:

---

*With or without pretreatment such as filtration, pH adjustment, coagulation, precipitation, flocculation, sedimentation, stabilization, thickening, and dewatering (Adams, 1991)

1. Biological oxidation/conversion (aerobic and anaerobic systems)
2. Chemical oxidation (air oxidation, ozonation with or without UV catalysis, permanganate with or without UV catalysis, chlorine dioxide)
3. Physical separation (coagulation/precipitation, gravity separation, dissolved air flotation, hydrosieve/microstrainer, granular media filtration, air stripping)
4. Chemical/catalytic conversion (hydrolysis, catalytic dechlorination, reductive dechlorination, photolysis)
5. Adsorption/exchange (activated carbon, macroreticular resins, ion exchange, natural polymeric, solvent extraction)
6. Concentration/volume reduction (steam stripping, vacuum distillation, evaporation/crystallization, freeze crystallization)
7. Membrane process (reverse osmosis, electrodialysis)
8. Thermal/catalytic destruction (incineration, solar destruction)
9. Other (chemical fixation, microwave plasma destruction, land disposal)

From these brief technology descriptors, many classic safety concerns are evident such as corrosivity, solvent toxicity, heat, cold, toxic gases and vapors, metal toxicity, confined spaces, rotating equipment, drowning, radioactivity, and other hazards. Transportation of hazardous waste materials is also a related safety concern, but will be treated in the Transportation section.

Waste handling operations deserve a comprehensive system safety analysis to recognize, evaluate, and control a wide variety of hazards and associated risks.

WELDING SAFETY

Welding can involve many different types of processes. Most common is electric arc and gas (fuel gas/oxygen) welding, but there are many variations, such as inductive welding, percussive welding, flash welding, plasma welding, sonic welding, thermite welding, and others. Arc welding variants include gas shielded welding, metal arc welding, gas-tungsten arc welding, gas-metal arc welding, flux cored arc welding (McElroy, 1980).

An OSHA study of 200 welding/cutting-related deaths over an eleven-year span showed 80% lack of adherence to safe procedures. Eleven percent of deaths involved malfunctioning/failed equipment and only four percent were related to environmental factors (*Professional Safety*, Feb. 1989). This strongly suggests that the basic equipment design hazard causes have been relatively well controlled and that safety engineers should only have to confirm that equipment has not been modified or maintained adequately. Then the emphasis should shift to safe operations.

Welding ignition of flammable materials is a common causal factor in fires and explosions. In addition to acting as an ignition source and possibly the flammable gas source, welding causes eye injuries from ultraviolet radiation overexposure and foreign bodies, heat stress illness and burns, excessive noise for some operations such as arcjet cutting, potential high-pressure system failures, manual handling injuries from heavy components or cylinders, electrical shock, toxic fume illness, and asphyxiation potentials. Laser welding adds a hazard dimension in control of the intense laser beam and its reflections. Additionally, associated weld grinding and rotary wire wheel brushing cause serious injuries.

Fire/explosion hazard controls include backflash devices in hoses and at tanks, containment of welding slag, stringent flammable material exclusion controls, flammable mixture monitoring and alarm, and specialized fire protection measures.

Eye protection centers on protective eyewear, both for ultraviolet and laser radiation, along with face/eye protection for flying foreign bodies such as wire brush and weld ejecta. It should be cautioned that laser goggles, in particular, are good for narrow ranges of wavelengths.

A particularly attractive development from a productivity standpoint is the clear-view type welding helmet (e.g., Winkin or Hornell Speedglas) which automatically darkens when an arc is struck, avoiding the need to raise and lower the hood when changing rods or weld location.

Heat stress is usually controlled with ventilation controls, cooling garments, work rotation/limitation, and fluid intake.

Burn protection is offered by personal protective clothing. Besides the old standby leather and impregnated cottons, many new fire-resistant fibers have been developed to take the place of asbestos.

Excessive noise is controlled by special welding helmets which offer hardhat protection, ventilation, radiation, and noise protection in one device, and a variety of noise personal protective equipment.

High-pressure failures of welding systems are generally controlled by front-end specifications and by nondestructive testing, hydrostatic proof testing, and visual inspection. However, it is necessary to control cylinder usage to avoid common mishaps such as knocking off a cylinder valve, damaging a connector, line, or regulator.

Manual materials handling (MMH) of heavy gas cylinders and heavy materials being welded cause many strains and some serious injuries. Material handling devices and sufficient manpower are excellent methods of improving MMH.

Electrical shock from electrical welding machines is not usually catastrophic unless it causes a worker to fall. There is a broader discussion of the latter in the electrical safety chapter. Good grounding, bonding, and insulation are usually sufficient to avoid serious electrical shock.

Toxicity hazards are best controlled by use of low toxicity materials (such as in rods and fluxes), local exhaust and area ventilation controls, and respiratory protection, depending upon the type and level of toxicity. The amount of fume produced can be minimized by careful control of welding machine parameters and welding technique. Welding fume fever (from welding on zinc-containing materials) plus chrome and lead poisoning (from removal of lead chromate paint) are common toxicity concerns with welding.

Asphyxiation associated with welding comes from undetected inert gas accumulations, depletion of available oxygen in confined spaces, both flash fire and partial oxidations forming carbon monoxide.

5

SAFETY MANAGEMENT FUNCTIONS AND PROGRAMS

Management was not often applied to safety in the past, probably due to low esteem for safety, low emphasis, and lack of promotion potential for managers, among other reasons. Early safety programs focused on injuries and were performed by someone on the shop floor, who may often have been injured. Many of these people were sincere, mechanically apt, but ill prepared to manage. Since many thought that safety was just a matter of a bad attitude or "carelessness," who better to make responsible for safety?

The underlying reason for managing safety is that the consequences of not achieving a satisfactory level of safety may threaten the viability of the business. Management is needed to plan, organize, lead, and control the achievement of safety.

Safety management is proposed as a science developed on logic, strategy, and tactics of administration, using principles and concepts rather than safety rules and regulations. Management does not take to safety because it perceives safety to be unrelated to the basic goals of the enterprise. This attitude strongly suggests that safety must be integrated into the primary functions of a business, including business goals.

Management failures have increasingly been blamed for poor safety performance. If management is responsible for organizational performance, including safety performance, there is credibility to the assertion. Whether all blame belongs to management failures is much less certain. Nevertheless, management's role is getting a lot more scrutiny. William G. Johnson, in his 1980 book *MORT Safety Assurance Systems,* focuses management oversights and omissions, assumed risks and general management system weaknesses. The National Safety Management Society has been outspoken in addressing management responsibilities and approaches in safety risk resolution.

MANAGEMENT FUNCTIONS

Management means getting things done through others, with a minimum of error. Work injury and property damage are thus tangible evidence of bad management. Modern safety management must not stop at the identification of unsafe acts and conditions. It must be competent in the management of its resources and the accomplishment of results through others. Managerial skill is a complementary skill to technical engineering skills.

The breakdown of the four main classifications of management activity, given in Figure 5-1, shows activities directly applicable to safety programs.

Keep in mind that there is overlap between most management functions. Breakdowns are helpful to focus on individual subfunctions, but these subdivisions are not exclusive or independent in the mathematical sense. Remember that it is always necessary to return to an integrated viewpoint in the end. Safety management will require decomposition and integration in a classic analysis approach.

Safety management planning requires the ability to forecast loss, costs, and trends. Despite short-term inability to accurately predict losses, good histories and trend information can give businesses enough information to plan for safety-related costs. Proactive planning, rather than reactive responding, is needed in safety. Planning must cope with the change that is increasingly

PLANNING	ORGANIZING
Forecasting Setting Objectives Formulating Policies Budgeting Determining Procedures	Structure Delegation Relationships
LEADING	CONTROLLING
Deciding Motivating Communicating Selecting Developing	Setting Standards Measuring Performance Evaluating Performance Correcting Performance

FIGURE 5-1 Breakdown of the Four Main Classifications of Management Activity

required of all businesses. Safety objectives are necessary in the larger set of corporate objectives and at the unit level. Safety objectives need to be concrete, attainable, and measurable. Safety policies are likewise necessary for consistency and direction. Programming refers to the procedural steps necessary to accomplish goals, their organization, including the necessary provisions and interactions. Budgets are a measure of organizational independence, emphasis, and scope of activities through resource and authority limits. Safety programs without a budget are usually deemphasized, dependent, and organizationally weak. Determination of procedures is a well-recognized safety planning function, that normally requires analysis rather than being done a priori.

Safety management organizing requires staffing structures, teamwork and, in today's participative structures, sharing and delegation of functions. Safety plans must indicate the relational frameworks for safety organizations.

Safety management leaders are often staff advisors to line management. Decisions are made within the safety organization but the much larger influence must be accomplished with the line organizations. Dotted line relationships (matrix relationships rather than direct reporting) are often used to share limited safety expertise throughout organizations. Recently, with organizational downsizing and streamlining economy moves, self-directed work teams are replacing many line and staff supervisory personnel. This means that safety management numbers would shrink and it also shifts program emphasis to non-management individuals. The previously well-known National Safety Council's "Key Man" program concept, which places the greatest emphasis upon the first line supervisor, would require another point of application if the supervisor has been eliminated.

Motivation is a difficult skill to master and use in the accomplishment of results. Robert Townsend's insights to motivation are helpful. He says, "Don't change people, just utilize their skills to advantage. The motivational door is locked from the inside—create a climate for the person to motivate himself to reach corporate goals."

Communication is a routine safety function. Proper selection and development of staff both are necessary, especially where matrix management is used. Matrix management requires personnel to answer to more than one boss and, in some cases, to persuade superiors to provide resources.

Safety management controlling is a traditional area of safety strength in many organizations. Much emphasis is placed on the setting of safety standards, measuring and evaluating organizational performance against those standards, and usually recommending corrective actions to line management for deficient performance.

FUNCTIONS OF THE SAFETY PROFESSIONAL

Safety professionals are closely aligned with the functions of safety management. The American Society of Safety Engineers' (ASSE) "Scope and Functions of the Professional Safety Position," categorizes the functions of safety professionals as indicated in Figure 5-2.

Dan Peterson summarized those four areas as follows:

1. Systems to identify deficiencies
2. Program development—approaches and plans to correct deficiencies
3. Communication—transmission of plans of action
4. Measurement—valid determination of plan performance

It is helpful to examine the definition of professional to understand what ASSE is driving at in their discussion for safety engineers. Webster's *Third New International* (unabridged) describes professional as

One of the *learned professionals* characterized by a *high level of training and proficiency . . .* requiring *specialized knowledge* and often long and *intensive preparation,* including *instruction in skills and methods,* as well as *scientific,* historical or scholarly skills which underlie such skills and methods. The profession is *maintained by force of organized or concerted opinion high standards of achievement and conduct,* and committing its members to *continued study* and to a kind work which has for its prime purpose the *rendering of a public service* (my emphasis).

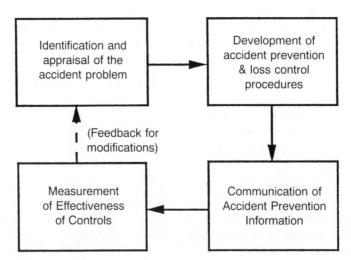

FIGURE 5-2   Functions of the Professional Safety Position   *Source:* ASSE.

I believe safety engineers should be professionals and this applies to those who qualify for the Certified Safety Professional (CSP) designation, offered by the Board of Certified Safety Professionals. The certification is similar to that required of professional engineers. Some states allow reciprocity, with regard to qualification, to CSPs for safety engineering practice in their state.

Douglas Hardie has described the underlying health and safety task with the diagram in Figure 5-3. He further gives a helpful adaptive model of safety practitioner roles and activities in Figure 5-4.

In this figure he suggests three levels of qualifications of safety practitioners: (1) Expert/problem solver/doer, (2) Questioner/researcher, and (3) Facilitator. By his descriptions of these levels, Hardie considers practitioners to be below the level of managers. It indicates what the staffs below safety managers generally do.

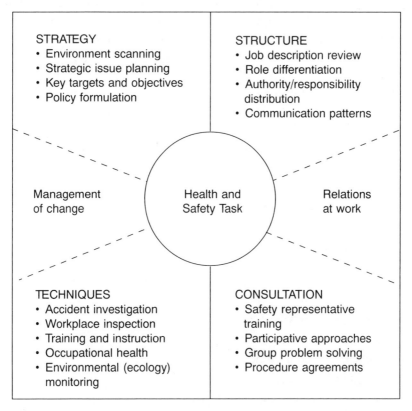

FIGURE 5-3   Underlying Health and Safety Task *Source:* Hardie, D. (1981) "The Safety Professional of the Future," *Professional Safety.*

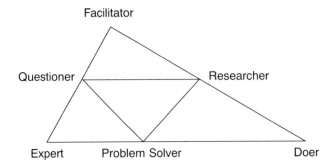

FIGURE 5-4 Adaptive Model of Safety Practitioner Roles and Activities

SAFETY PROGRAMS

Programs are organized actions involving defined activities and resources taken to reach an objective within a given time. Management makes use of programs to achieve its objectives. The elements and examples showing the relative scope of safety programs helpful to grasp the program concept, are given below.

The elements of safety programs are many, but include the following:

- Understanding and competence
- Participation and practice
- Placement of responsibility
- Consulting versus doing
- Monitoring of performance
- Integration of objectives
- Proper point of control

A minimal safety program contains a few elements like these:

- Preplacement physical exams
- Management enforcement of policy/procedure
- Productive training program
- Preventative maintenance
- Accident reporting and investigation

A basic outline of a comprehensive safety program follows:

I. Policy
   A. Management expectations
   B. Establishment of a safety and health system
II. Organization
   A. Structure
   B. Staff
   C. Communications
   D. Reporting
   E. Budget
   F. Facilities and equipment
III. Safety and Health Plan
   A. Inventory of operations
   B. Safety Engineering
      1. Hazard identification
      2. Assessment of risk
      3. Hazard and risk control
   C. Risk acceptance
   D. Safety and health training
   E. Program evaluation
   F. Record keeping
   G. Statute compliance
   H. Emergency response

Issuing a safety manual is a common way for organizations to set out safety policy, safety requirements, and safety procedures. Although common, a safety manual is not the only or the best way to delineate program directives. For instance, directives can be integrated into operational manuals which are readily available for use at the job site.

Note that safety policy and safety procedures are different. Safety policy is a strategic "where are we going" statement of direction, whereas safety procedure is more a tactical "how to" set of instructions to accomplish the policy.

Insight into programs is often gained by reviewing their safety manual's table of contents. No one arrangement of topics is best (there is no universal standard for program organization). Topics need to be tailored to the program. Example headings adapted from two safety manuals are given below:

Sample safety manual headings, example 1

- Policy and purpose
- Administration
- Standards

- Mishap notification, investigation, reporting
- Inspection and abatement
- Safety training and awareness
- Safety committees and employee rights and responsibilities
- Medical treatment and first aid
- Personal certification for hazardous operations
- System safety and risk assessment
- Operational safety
- Contractor operations
- Safety program evaluation and review
- Public safety
- Operating safety requirements
  —Confined space entry
  —Construction safety
  —Cryogenic materials
  —Electrical safety
  —Explosive and propellant safety
  —Emergency action
  —Facility safety
  —Fire safety
  —Hazardous materials
  —Hazardous operations
  —Liquids and gases
  —Materials handling
  —Office safety
  —Personal protective equipment
  —Pressure systems
  —Underwater operations

Sample safety manual headings, example 2

- Safety management and administration
- Accident investigation, reporting, and analysis
- Planning for safety
- Personal protection
- Hazardous materials

- Storage, use, and handling of hazardous materials
- Lab safety
- Construction and maintenance
- Motor vehicle and transportation equipment
- Industrial hygiene
- Fire prevention and protection
- General safety requirements

Statements of safety policy are important to demonstrate management resolve and management support of safety objectives. Input from management and even boards of directors is necessary in addition to safety organization input before drafting or revising safety policy.

Too often, policies are relatively worthless and too innocuous to be of any value. I believe that safety policies should use similar approaches to other important business function policies. Niru Davé (1989) identifies three components of a safety and health policy: Management's concern and interest, Direction, and Expectations. He also discusses the need for proper approval and implementation. No policy can be expected to stand indefinitely, so regular review is important.

In my opinion, a safety and health policy should call for commitment to excellence rather than mere compliance with minimum legal requirements. Expected results should be measurable rather than appealing to emotion or political correctness. Objectives should be tied to corporate values, which would include being a responsible corporate citizen, better than average safe working conditions, effective return on investments, and so on. Implementation of the policy should be addressed, including who does it, when, and how the implementation is assured. It is acceptable to point to implementing directives in the policy, so long as the objectives have been summarized.

The following are pitfalls to avoid in policy statements:

1. Vagueness to extreme, especially motherhood statements, cliches, and open-ended challenges
2. Confusing statements
3. Contradictive statements
4. Negative tone or language with overtones
5. No assignment of responsibility, no gatekeeper
6. Overly complicated language or "legaleze"; use of technical jargon where lay terminology would do
7. Inflexibility

OSHA, in its "Managing Worker Safety and Health" guidebook, gives the following (adapted) advice on writing safety and health objectives for a six-to-12-sentence safety and health policy statement:

- Start with an action verb
- Specify a single key result to be accomplished
- Specify a target date for accomplishment
- Be specific and quantifiable (measurable and verifiable); specify what and when; avoid why and how
- Relate directly to the accountable manager's role
- Be realistic and attainable, yet present a significant challenge
- Provide a maximum payoff for the investment of time and resources

Implementation of safety and health policies may be done with management representatives or in the absence of mid- or first-line management, whoever accomplishes the management functions. A common management practice for implementation of safety policy is the use of the safety and health committee or "safety committee." In some cases this has been mandated by management–worker contracts. For the safety committee to be truly productive, it must be empowered and allowed to manage its own destiny to a point. Charters and scope of responsibility for safety committees must fit with the safety objectives for the entire organization. Management techniques should be applied within the safety committee, such as objectives, methods (especially group problem solving), schedules, deadlines, agendas, minutes, and reporting.

CONCLUSION

In my view, we have safety engineering on the right, behavioral safety advocates on the left, and many shades in between. Much of the difference is muddied by definitions. Safety management is not so much a separate discipline as it is an emphasis to be laid on top of organizational systems to deal with safety. I strongly support safety engineers incorporating sound management approaches into the practice of safety engineering. Detailed scientific behavioral approaches are considered outside the scope of this text.

6
SAFETY DOCUMENTATION

Safety documentation is the necessary recording of safety information. It includes safety analysis reports, safety assessments, safety investigation reports, safety audit reports, safety surveys, safety statistics, safety plans, safety procedures, problem reporting, and so on. The adjective *safety* has been included in the previous list of documentation because it is commonly associated with specific documentation.

In this text, safety documentation does not refer to safety literature, safety catalogs, safety books, and all sorts of general safety references (discussed in the next section). Here, safety documentation applies to the systematic documentation of program actions (what actions are taken by whom at what time, what resulted, and what is further needed). Safety documentation also includes listing the sources of loss information, insurance statistics and loss ratings, medical reports, epidemiological studies, industry studies, trade studies, and so on. Systematic documentation suffers from some unavoidable overlap between categories, and therefore requires that ground rules be explicitly stated.

SPECIFIC SAFETY DOCUMENTATION

There is a trend in OSHA standards to require specific safety documentation. This requirement may be somewhat reduced by the trend toward "performance standards," which are discussed under safety references in the next section; however, some documentation will be necessary to be able to demonstrate adequate program performance.

In more formalized safety programs, safety documentation requirements are given in contracts, often as "Safety Data Requirements (DRs)." DRs normally give some schedule and frequency requirements, format, routing of copies and, in some cases, provide references to analytical methods and other "how to" implementation directives.

Because safety documentation of performance can be used to demonstrate corporate intent, compliance, and avoid larger legal penalties such as for so-called "egregious" violations of laws, the following types of safety documentation are examples of potentially valuable programmatic information:

- Lists of requirements which apply to company operations (at the time), including all revisions and dates of such requirements in use
- Safety and health policy statements, including handbooks and manuals, safety data sheets (intermediates and products), and safety instructions for products
- Chemical, physical, and ergonomic hazards per job, including exposure records (such as job-exposure histories) by person (to include employees and contractors
- Analyses of job risk, including hazard identification, hazard evaluation, hazard control, and safety verification, including analysis dates
- Records of specific complaints, including claimant, date, and allegations
- Records of legal citations, abatements, awards, judgments, and so on
- Records of any safety and health-related significant events, including catastrophic explosions, chemical releases, process failures, and so on.

Safety documentation is particularly valuable as a record that may be introduced into court proceedings, especially for long-past activities. It is important that record-retention policies be carefully reviewed by safety and legal departments. From the lawyer's standpoint, safety documentation is a two-edged sword, and legal guidance would be in order regarding the content and procedures for documentation. Stated plainly, records can show good performance and bad; compliant or noncompliant activity. Documentation can help defend or convict, depending upon content. The best policy may be to expect to have to divulge all safety documentation in the future. It is the organization's responsibility to show long-term correction of shortcomings, which might be indicated in short-term documentation.

ANALYSIS OF DOCUMENTATION

Since safety analyses depend upon methodology assumptions and rationale used to draw conclusions, safety organizations should specify the minimum content for documentation of specific analyses. For instance, fault tree analysis should contain generous amounts of documentation to explain the anal-

Accident	Accidental	Accident Prevention*
Alarm		
Catastrophe	Catastrophic	
Confined Space*		
Cumulative Trauma Disorder	CTD	CTS
Danger	Dangerous	
Death		
Design Safety*		
Disaster		
Electromagnetic Interference	Electromagnetic Compatibility	
Emergency	Emergency Procedures*	Emergency Response
Energy Exchange*		
Ergonomics*		
Explosion	Explosive	
Failure Modes and Effects	Failure	FMECA*
Fall	Fall Restraint	
Fatal	Fatality	
Fault	Fault Tree	FTA*
Fire	Fire Safety*	
First Aid		
Guard	Guarding	
Harm	Harmful	
Hazard	Hazardous	Hazard Level*
Hazard Analysis	HA*	HR*
Hazard Control*	Hazard Control Verification*	
Hazardous Materials*		
Hazardous Processes*		
Hazardous Waste		
Hearing Loss		
Human Error		
Illness		
Industrial Hygiene*		
Inhibit		
Injury		
Lockout	Lockout/Tagout*	
Logic Tree		
Loss	Loss Control*	Loss Prevention*
Noise	Noise Control	
Occupational Hazard*	Occupational Safety*	
Occupational Health	Occupational Health & Safety*	OH&S*
Off-the-Job Safety*		
OSHA*	NIOSH*	OSHREC*
Product Safety*		
Propellant Safety*		
Pyrotechnic Safety*	S&A*	
Risk	Risk Assessment*	Risk Acceptance

FIGURE 6-1   List of Safety-Related Keywords (*continued on next page*)

Safety Restraint*	Safety Belt*	Safety Harness*
Safety Certification*		
Safety Codes*	Safety Regulations*	Safety Standards*
Safety Consulting*		
Safety Interlock*		
Safety Monitoring*		
Safety Professional*		
Safety Statistics*		
Safety Training*		
Sneak Analysis*	Sneak Circuit Analysis*	
Toxic	Toxicity	
Unhealthy		
Unsafe		
Warning	Caution	
* Strongest Association		

FIGURE 6-1 List of Safety-Related Keywords (*continued*)

ysis to higher management, to simplify the next iteration, and to guard against loss of "corporate" knowledge when the analyst is no longer with the organization, is no longer available, or has forgotten the basis for the analysis.

Databases should be utilized to track analyses and their information. For instance, a database of hazard report's hazardous conditions, causes, controls, and verifications, is very helpful for future analyses. If a good key-word index is developed along with the database entry, it is possible to work toward an expert system or at least simplify the analyst's research. (See the suggested key-word list in Figure 6-1.)

Databases of safety requirements are available. Some may be offered commercially for specialized areas, such as aerospace. Due to rapid changes, these databases probably should be on-line. Also, due to the incredible volume of requirements (some overlapping and some contradictory), there needs to be a multi-tier arrangement of data. This arrangement would allow a researcher to search lower levels, if desired, without incurring the costs of searching the entire data set. In this way, requirements and related requirements could be summarized and cross-linked. An example of a lower level would be actual text of a requirement. One lower level could be interpretations of those requirements. A similar level could be actual use of those requirements by a certain in-company organization.

CHANGE MANAGEMENT

A note of caution is in order concerning safety documentation. It may be very important to submit certain documentation via a change management (CM) system. Formal CM systems evolved to keep track of complex configurations such as weapons, and now are applied to rapidly changing consumer products, such as automobiles. CM may be required by contract, or it may become prudent to know what was done in a certain time frame. Too often, safety documentation is updated at will, and no record of previous documentation of system state is kept. A proper CM approach to safety documentation can also introduce requirements for those changes which affect safety to be identified during change to the safety organization(s). In particular, specifications, drawings, processes, and procedures may change often during a life cycle, and safety review may not occur if the reviewers are unaware of the configuration history.

7
SAFETY REFERENCES

Safety references abound. The trick is to know where safety information resides in peripheral sources. Experience is useful and so is a good bibliographical listing. The References at the end of the book has been created from my experience. It should give a starting place in a broad variety of safety-related topics. It is not suggested that this listing is either exhaustive or complete. Bibliography listings such as this change too quickly to be current and may need full-time attention to be properly maintained. The following table is a short working list of safety references.

Topical subject lists also help in identifying safety subjects. In these days of computerization and information searches, key words are of great value. Unfortunately, there are many potentially safety related key words. Searching on "safety" alone is not always helpful. "Safety" + "Engineering" better limits the field; however, most searches should be conducted with the particular skill area or topic narrowed as much as possible. "Chaining" from identified references to citations in their bibliographies is often helpful. Chaining is the linkage between a secondary, tertiary, and so on reference, back to the primary written source of the information.

There is a tremendous bulk of standards, codes, regulations, and requirements in print, many of which have been converted into microfilm, microfiche, optical disk, and various electronic database formats. It is generally impractical for any organization to hold all these references. Since there are so many of these references and they change quickly, electronic information services are blossoming.

COMPUTERIZATION

Due to the information explosion, it should not be long until safety engineers must use computer services to do their jobs efficiently. Each engineer will

List of Safety References

Title/Edition	Author/Editor	Publisher	Date
Assurance Technologies—Principles and Practices	Dev G. Raheja	McGraw-Hill	1990
Biohazards Management Handbook	Daniel F. Liberman and Judith G. Gordon, eds.	Marcel Dekker	1989
Chemical Process Safety: Fundamentals with Applications	D. A. Crowl and J. F. Louvar	Prentice-Hall	
Emergency Planning and Community Right-to-Know: An Implementer's Guide to SARA Title III	Victoria Cooper Mussleman	(ASSE distribution)	1989
Encyclopedia of Occupational Health and Safety, 3rd ed., 2 Vol.	International Labor Office, Geneva	(ASSE distribution)	1987
Engineering Design for the Control of Workplace Hazards	R. A. Wadden and P. A. Scheff	(ASSE distribution)	1987
Ergonomic Design People at Work, 2 Vol.	Eastman Kodak Company	Van Nostrand Reinhold	1983, 1986
Ergonomic Interventions to Prevent Musculoskeletal Injuries in Industry, Industrial Hygiene Series 2	American Conference of Governmental Industrial Hygienists	Lewis Publishers	1987
Ergonomics Sourcebook: A Guide to Human Factors Information	Kimberley H. Pelsma, ed.	(ASSE distribution)	1987
Eshbach's Handbook of Engineering Fundamentals, 4th ed.	Byron Tapley, ed.	(ASSE distribution)	1989
Fire Protection Handbook, 17th ed.	Harold E. Nelson, et al.	NFPA	1992
First Aid Manual for Chemical Accidents	Marc J. Lefevre	(ASSE distribution)	1980
Guide to Safe Practices in Chemical Laboratories	Royal Society of Chemistry	CRC Press	1986

Title/Edition	Author/Editor	Publisher	Date
Guidelines for Laboratory Design	Louis DiBerardinis, et al.	(ASSE distribution)	1987
Handbook of Hazardous Waste Regulation, Vol. 2: *How to Protect Employees during Environmental Incident Response*	EPA	(ASSE distribution)	1985
Handbook of Human Factors	Gavriel Salvendy, ed.	John Wiley & Sons	1987
Handbook of Occupational Safety and Health	Lawrence Slote	(ASSE distribution)	1987
Handbook of Rigging	W. E. Rossnagel	(ASSE distribution)	1988
Hawley's Condensed Chemical Dictionary, 11th ed.	Rev. by N. Irving Sax and Richard J. Lewis, Sr.	(ASSE distribution)	1987
Hazardous and Toxic Materials: Safe Handling and Disposal, 2d. ed.	Howard H. Fawcett	(ASSE distribution)	1988
Hazardous Chemicals Desk Reference	N. Irving Sax and Richard J. Lewis	(ASSE distribution)	1987
Health and Safety in Agriculture	James A. Dosman and Donald W. Cockcroft, eds.	CRC Press	1989
Human Factors in Engineering and Design	M. S. Sanders and E. J. McCormick	(ASSE distribution)	1987
Industrial Ventilation Workbook, 2d ed.	D. Jeff Burton	(ASSE distribution)	1990
Introduction to Fall Protection	J. Nigel Ellis	(ASSE distribution)	1988
Investigating Accidents with STEP	Kingsley Hendrick and Ludwig Benner, Jr.	(ASSE distribution)	1987
Major Industrial Hazards—Their Appraisal and Control	John Withers	Wiley-Interscience	1988

(continued)

List of Safety References (*Continued*)

Title/Edition	Author/Editor	Publisher	Date
Materials Handling Handbook	Raymond A Kulwiec, ed.	(ASSE distribution)	1985
McGraw-Hill Dictionary of Scientific and Technical	The McGraw-Hill staff	(ASSE distribution)	1987
Modern Accident Investigation	Ted S. Ferry	(ASSE distribution)	1985
National Fire Codes	NFPA	NFPA	1990
New Directions in Safety	Ted S. Ferry, ed.	(ASSE distribution)	1985
Occupational and Environmental Safety Engineering and Management	Hamid R. Kavianian and Charles A. Wentz, Jr.	(ASSE distribution)	1990
Occupational Safety Management	Willie Hammer	(ASSE distribution)	1985
Patty's Industrial Hygiene and Toxicology, 6 Vol.	George D. and Florence E. Clayton, eds. (Vols. 1, 2A, 2B, 2C); Lewis J. and Lester V. Crally, eds. (Vols. 3A and 3B)	(ASSE distribution)	Vol. 1: 1978; Vol. 2A: 1981; Vol. 2B: 1981; Vol. 2C: 1982; Vol. 3A: 1985; Vol. 3B: 1985
Pesticide Users' Health and Safety Handbook	Andrew Watterson	Van Nostrand Reinhold	1988
Photo Techniques for Accident Investigation	ASSE	(ASSE distribution)	1984
Practical Electrical Safety	D. C. Winburn	(ASSE distribution)	1988
Product Liability: Design and Manufacturing Defects	Lewis Bass	(ASSE distribution)	1987
Product Safety and Liability: A Desk Reference	J. Kolb and S. S. Ross	(ASSE distribution)	1979
Professional Liability of Architects and Engineers	Harrison Streeter	(ASSE distribution)	1988

Title/Edition	Author/Editor	Publisher	Date
Profitable Risk Control: The Winning Edge	William Allison	(ASSE distribution)	undated
Red Book on Transportation of Hazardous Materials, 2d ed.	Lawrence W. Bierlein	(ASSE distribution)	1987
Safe Storage of Laboratory Chemicals	David A. Pipitone, ed.	(ASSE distribution)	1984
Safety and Health for Engineers	Roger L. Brauer	(ASSE distribution)	1990
Safety in the Use of Industrial Robots, OS&H #60	ILO	ILO Publications	1989
Safety Manual for Municipalities	David A. Dodge	(ASSE distribution)	1986
Safety Training Methods	J. B. Revelle	(ASSE distribution)	1980
Sax's Dangerous Properties of Industrial Materials, 8th ed.	Richard Lewis, Sr.	Van Nostrand Reinhold	1992
System Safety Engineering and Management	H. E. Roland and B. Moriarity	(ASSE distribution)	1983
The Practice and Management of Industrial Ergonomics	David C. Alexander	(ASSE distribution)	1986
The Safe Handling of Chemicals in Industry, 2 Vol.	Phillip A. Carson and C. J. Mumford	Wiley-Interscience	1989
Training in the Workplace	Earl Heath and Ted Ferry	(ASSE distribution)	1990
Trench Safety Shoring Manual	Red Cass and Matt R. Wall	(ASSE distribution)	1989
The Wiley Engineer's Desk Reference	Sanford I. Heisler	(ASSE distribution)	1984
VDT Health and Safety Issues and Solutions	Elizabeth A. Scalet	(ASSE distribution)	1987

have a personal computer for doing information searches into requirements and guidance documentation, eventually via the "information highway" which will connect various information sources. Costs of finding information should decrease through the information highway, although there will continue to be tolls just as in transportation highways. The ability to tie recommendations to existing documentation will lend credibility and aid standardization. Increased computer capabilities will allow better utilization of data through data compression, data organization, interaction between data formats (text, graphics, video, sound, and even virtual reality), and the previously mentioned data searching methodologies.

ELECTRONIC DATABASES

The government is establishing many electronic databases, some of which have their focus in safety and health. Additional rapid growth is occurring in the commercial database business. The more enlightened ones scan and translate as necessary, using extensive key word systems. With the establishment of linkage systems such as Apple's HyperCard™, self-customized databases are becoming a simpler reality within the reach of most.

SAFETY TEXTS

The number of comprehensive safety texts is still relatively small, considering the number of professionals and the age of the discipline. However, this situation is rapidly changing per the general information growth trends. Safety handbooks are of assistance to the neophyte safety engineer. Topics vary between handbooks because of lexicon differences and the degree of art involved. Safety specialization is best treated in individual texts by experts.

SAFETY JOURNALS

Safety journals represent another good source of information, but not all information in safety journals is well indexed. Associations and society members publish through these journals. In my opinion, journals tend to be a cut above other publications because peer review, as a part of the journal article acceptance process, tends to eliminate unsubstantiated or conjectural writings. Depending on the bias of the reviewers, some radical yet correct positions may be denied. Nevertheless, peer-reviewed articles are generally more trustworthy than unreviewed articles.

TRADE MAGAZINES AND DIGESTS

Trade magazines and periodical digests often contain timely articles about new products and legislative developments. Most of these articles are written for a more casual audience than the journals address. My appraisal of such articles is that recently, quality is generally converging upward, toward the high standards of most journals.

SAFETY DATABANKS

Safety research databanks exist in many forms. These range from sophisticated sets, like human error rate information, to simpler sets, like recommended weight limitations for rigging components. Databanks must be carefully scrutinized for proper application and veracity but are better than an information void.

SAFETY APOCRYPHA

A closing warning about safety apocrypha is in order. Logical sounding and oft-quoted statements do not necessarily ensure veracity. Beauty is in the eye of the beholder, and closely held beliefs often get mixed in with facts. The viewpoint and background of any observer color the interpretation of what is seen. Also, misinformation tends to get quoted over time. Examples of safety apocrypha include "Over 85% of all accidents are due to operator error" and "Awareness of the hazard will correct safety hazards." Safety engineers need to be vigilant for safety apocrypha.

8

CASE STUDIES OF SAFETY ENGINEERING APPLICATIONS

Case studies are a way to apply safety engineering principles to situations that resemble real life. The following case studies were drawn from my experience and modified to bring out and emphasize certain information (and to protect the innocent). Each study is described, then discussed further (such as investigation would provide), and finally, some generic answers are suggested. These examples are rather open ended, as true situations tend to be. The "answers" are those which could be formulated in an hour or two. Detailed analyses could take much longer. The case studies may be useful for group evaluation, with the instructor (in a role-playing mode) providing additional hypothetical details, as needed by the teams. Teams could summarize their findings and report to the entire group. This would show the normal variation in problem-solving solutions, and would illustrate the normal safety engineering function of reporting findings and presenting results.

CASE I

Case I involves a corporate banking system coin operation. Vibrating inclined wheel sorters are used to sort large numbers of coins, which are fed into wrappers and stacked into boxes which weigh from 25 to 35 pounds each when full. The boxes are stacked on wooden shelves which sit on the floor and form a counter top. There is a base and a middle shelf, and the depth of the shelves is about 36 inches. Boxes are moved by stacking onto metal carts which are pulled or pushed to bank locations on upper floors.

The workers are all quite junior men, with an age of 20+/−2 years. The longest tenure on the job is 18 months. Several former workers suffered back injuries, but none of them are still employed because they were fired for being incapable of performing the work. Other former workers suffered an unusual rash on their hands.

A noise level of 92 dBA was obtained when the coin sorters are operating. No further measurements were made because of possible liability for hearing claims. Sorting is pretty much continuous during all dayshifts. No hearing protection is required and none is worn.

The bank would like to have more coin availability, but is limited by the number of bonded, semi-skilled workers. A higher bank efficiency expert suggests that incentives or performance requirements be placed on workers.

There are no written task instructions, in part because many of the employees do not read well. There seems to be quite a variation in how the tasks are performed.

Discussion

This case contains a number of opportunities for improvement, as safety optimists are prone to say. Many potential risks have resulted in losses and others that surely ought to result in losses. Manual material handling is a big part of this job. Coin boxes are fairly heavy and do not have any handles, resulting in high grip strength requirements. Sometimes, workers have been known to stack two and three boxes for the transfer motion from waist level down to the shelves. Shelves are located so that workers twist and lower the load in order to stack the boxes. Placement at the back of the shelves requires bending at the waist and reaching a long distance before releasing the boxes. There is a similar problem when unloading the shelves onto the carts.

The transfer carts have swivel casters on all four corners. The carts in use now are pretty heavy because of bad experiences with carts buckling under the load. There is a story about thousands of coins being spilled once when the whole load tipped over. There are several ramps and a couple of tracks for security gates that chew up the wheels. The maintenance men have tried using steel wheels but the sorter personnel have complained about not being able to make the turns at high speed. Speed is said to be required because of the high demand for coins.

A couple of the workers, who refused to be identified, said that their ears ring and they couldn't hear properly until several hours after working.

Study into industrial hygiene books led to a discovery that nickel coins may cause contact dermatitis in certain individuals.

No one could recall OSHA ever inspecting this bank's facilities. Since there are no safety problems, no safety responsibilities have been assigned.

Management felt that banking is so safe, making a big fuss over safety would just "create" problems that "really aren't there."

CASE II

Case II involves a maintenance shop in a golf course complex. The course has 54 holes and there are extensive grounds around the country club. The shop services a variety of vehicles, including 3-yard dump trucks, a trash compactor truck, a road runner vacuum truck to empty septic tanks, a fire truck, pickup trucks, sedans, roller mowers, rotary mowers, tractors, a forklift, a large backhoe, and electric golf carts. The shop does tire work, including split rim tires, carburetor overhauls, brake jobs, and routine motor repairs such as radiator jobs and and water pumps. Minor paint touch-ups are done, using electrostatic paint guns. Groundskeepers work out of the shop, mowing, spraying and spreading insecticides and herbicides on the greens and ornamentals located around the complex.

Discussion

The trash compactor truck is aging and routinely gets the ram plate stuck midway because the edge of the plate is worn and tree limbs get wedged between the plate and the lining. The worker often climbs into the compactor section to clear jams using a crowbar. When he is ready for the driver to try the ram again, he knocks twice with the crowbar. There have been a few times that fires have occurred in the trash. Once some noxious smoke resulted, probably from all the outdated pesticide/herbicide cans that were being disposed of in the facility dump next to the lake that leads to the nearby river. The maintenance people also double as local fire fighters, so the fire was put out after about fifteen minutes. There was some question at the time whether the arsenicals were a problem. Due to the urgency, the fire fighters put the fire out without donning any self-contained breathing apparatus.

The vacuum truck pumps its sewage into a lift station, which delivers effluent into a line to the nearest town's sewage treatment plant. Occasionally, a maintenance worker is dispatched out to the lift station to climb down the rungs about twenty feet to where he can see if the pump is working. Some of the workers say they use a disposable paper filter mask because of the rotten egg odor.

Most of the work with the pickups and sedans is done using caster-type hydraulic jacks. Occasionally a pickup has a load of sod in the bed, because it is too much trouble to offload it and reload it since many jobs only last three or four hours. Usually, brake jobs are done working right off the jacks. The same is true of muffler/tailpipe jobs. The shop gets stuffy during the winter when jobs like adjusting the timing are done with engines running and all the roll-up doors down. When asked, several workers said they have splitting headaches on those days. Also, one worker, who asked not to be named, said that a pickup fell off a jack and smashed the scooter board that he had been using to cut off a muffler and tailpipe. Fortunately, he was taking a smoke break, blowing out a brake drum with compressed air, when the incident occurred. When the occasional motor has to be pulled, such as for "government" work on one of the employee's personal autos, the forklift is pressed into service. A chain is wrapped around one of the forks and then the engine.

The mowers must be tested by maintenance workers once repairs are made. This is done by taking a high speed run on the edge of the course not too far from the shop. It was disclosed that several of the mechanics have been thrown from mowers. Most had driven tractors on nearby farms, so they thought they knew how to run them. There are no seat belts on any of the large riding mowers which mow up to a twelve-foot width in one swath.

Many of the radiators are able to be fixed by some brazing and soldering. This interferes with painting, which is also done inside to keep the dust from wrecking the finish. Sometimes, the mechanics doing the painting get zapped from a charge on the gun. Once, shop people welding set some masking newspapers on fire. It was touch-and-go to keep the wood shavings cleaning compound from catching fire because the compound is left on the floor to quickly absorb all the spilled crankcase oil. It could have been a costly fire since 50 fiberglass electric golf carts are charged right outside the shop wall. Battery servicing and replacement is done in the charging area. It is about 35 feet from the nearest entrance and about that far to the one restroom/washroom. Batteries are heavy duty, and weigh about 50 pounds each. The shop foreman stopped letting people use company coveralls because the acid kept "eating holes in them."

The shop has a rig for breaking down and reinstalling radial tires on rims. One mechanic said that it was nice having enough pressure to fix truck tires, as well as smaller tires.

The "chemicals" storage shed is attached to the shop, on the other end from the battery charging. Lots of things "you can't get anymore" are kept because the former head greenskeeper had a lot of pet treatments for every kind of grass ailment. When he died three years ago, nobody was exactly sure what was there. Most of the labels are missing. There is a lot of chemical residue on the floors. The greenskeepers are quite proud to be the only group within the shop to work with chemicals. There is a lot of turnover in this group

among the younger workers, but there are three greenskeepers over age 60.
Two of these workers have developed leukemia.

CASE III

Case III involves a commercial parking garage operation in a downtown
location. The garage has five floors and a couple of basement floors, and
covers a city block. Three floors are open to the public and the other floors
offer parking for a company headquarters. The company operates a large van
pooling operation and also has company pool cars.

One morning approaching eight o'clock, a customer reported a stream of
water flowing down the up ramp leading to the fifth floor. He also said that
there was a strong smell of gasoline. One of the parking personnel called
down to the sub-basement supervisor's office and they cut off the pumps,
which were being used to refuel vans. The building was evacuated and the fire
department spent the next 3 hours hosing gasoline to the sanitary sewer with
copious amounts of fire water. No fire or explosion resulted, so the parking
operation executive declared that there had been no accident.

It seems that a contractor was removing a pumping station from the fifth
floor that was no longer used. A gasoline supply line had been disconnected
over the weekend, but not capped off. According to parking officials, "about
300 gallons of gasoline" may have been lost.

Discussion

Upon checking, it was determined that no formal directions were issued for
the job. No drawings, no procedures, and no requirements were documented.
The workers were general laborers who spoke only Spanish, since they were
in the United States on a work permit. The contractor, who has performed
service station work in the past, said that these men obviously didn't understand their supervisor, or were basically careless.

It also seems that there had been two other "major spills of gasoline" in the
last six months. One involved overfilling a basement storage tank, when the
delivery driver dropped an extra hundred gallons or so into the street level
delivery pipe. Another spill was related to a leak in another sub-basement
storage tank.

The president of the division over the parking operations decreed that there
be no more gasoline spills, as it was embarrassing to have to explain them to
the corporate CEO.

SUGGESTED SAFETY ENGINEERING APPLICATION ANSWERS TO THE CASE STUDIES

Case I Answers

Serious job redesign is in order here. The fact that workers are voting with their feet is a clue that the job is a real "bear." Injury records must be reviewed and kept according to OSHA record-keeping requirements. Personnel practices regarding termination of injured workers should be scrutinized for compliance with existing laws. Turnover is costing the company dearly. It cannot afford to keep running through and injuring workers. Lawsuits are inevitable and awards may be massive, especially for permanent disabilities to young workers.

First of all, there should be a study done of lifting demands. The NIOSH Guide should be applied to determine which tasks are unreasonably risky. Work stations and work flow should be engineered to take out built-in twisting lifting and lowering. Conveyors, either gravity or powered types, may solve a lot of positioning problems. Containers should be redesigned to stay below recommended maximum weight limits. At the same time, containers may be designed to be easier to hold, such as by integrated container handles. Mechanical handling aids such as a lift table or pneumatic (suction) transfer arms may be applicable. Powered dollies may take the strain out of moving much larger loads than currently possible. Center of gravity should be lowered along with installing larger and improved wheels (such as with elastorper surfaces) on existing transfer carts to avoid tipping of loads. Obstacles and significant variations in elevation in floors must be eliminated. Carts should be pushed, rather than pulled, whenever possible.

Certainly there needs to be a matching of workers to the tasks, in terms of strength demands. Both static and dynamic testing should be reviewed for application. Conditioning of workers may pay dividends. Work practices need to be carefully specified in writing and incorporated into realistic training. Job safety analysis is an easy way to develop these methods, and should involve the workers in a group problem-solving effort.

Musculoskeletal strain is only a part of this case. Noise damage to hearing can be very expensive, if not controlled in a preventative manner. Tinnitus is often related to serious noise injuries. Noise level and dosage studies should be done with calibrated equipment, by a knowledgeable person. At the same time, audiometric testing of workers and prospective workers must be done to establish baselines and to detect threshold shifts. Depending upon results of noise studies, personal protective equipment such as foam earplugs and/or muffs should be provided and their proper use enforced, until engineering controls can be evaluated and introduced. Simple enclosures of coin sorters, lining of metal surfaces to deaden metal to metal impacts, use of viscoelastic dampening materials can be applied to control harmonic vibrations, and remote, reliable sorting is likely with more modern equipment. Preventative maintenance should be applied to gears, bearings, and closures to assure minimum vibration. Depending upon the manual need to manipulate coins, CTS may be likely. Wrist flexures and gripping requirements need to be reviewed for possible improvement.

Nickel allergies are possible, and are caused by contact with the nickel in coins. Suitable gloves are available, with types which allow for necessary dexterity. Caution is advised in their use where in-running nips exist that could pull in the fingers and hands by the glove. Patch testing of prospective workers could sensitize more people than would be allergic to nickel normally, so avoid this type of screening.

Case II Answers

This case involves several areas of safety concern. Obviously, a vehicular safety program is needed; but also there are quite a few operating hazards, so an operating hazard analysis would be in order. This analysis should identify and correct the unacceptable hazards associated with the operator clearing of jams in the compactor truck, as well as all the other operational hazards of other type motorized equipment.

The work with pesticides and herbicides is reason for safety analysis control of that type of hazards. The personal protective equipment that may be required will surely include respiratory protection equipment. A comprehensive respiratory protection program would include the evaluation of confined spaces, such as the sewage sump. The proper disposal of pesticides and herbicides in terms of safety and environmental concerns needs to be analyzed and corrected. Hazards to the public both from disposal and their normal use should be evaluated and controlled. Interface hazards to local fire fighters and other public service agency personnel need to be anticipated.

The garage work needs to be analyzed in terms of the energies, be it vehicles on jacks potential energy or tire work pressures, and toxicological concerns such as asbestos in brake drums or rusting mufflers. Materials handling involves common hazards in both the mechanical and manual lifting and positioning of objects. Ergonomic analyses of the manual aspects of lifting and the human-machine interactions with machines needs to be done.

The potential for work-related illness, especially cancer, should cause an epidemiologic study where cancer-causing or other health hazard chemicals are used, in spite of any personal protective equipment used.

The analyses suggested in these answers can be obtained with a combina-

tion of matrix and logic tree analyses. Regulations and programmatic requirements should show up in the controls that result from the matrix analyses.

Case III Answers

A series of near-misses, or near-hits, as the case may be, indicate strong
likelihood of unacceptable risk. Each incident should be evaluated with a
high-potential or similar analysis. The potential for catastrophe seems ripe
with underground storage of a high vapor pressure flammable liquid like
gasoline. When the public exposure—and running engines—are considered,
all that is needed is a vapor cloud from a leak, forming a flammable concentration in air, and then an explosion or flash fire results. If the multistory
building's structure is failed, fatalities and improbable rescues are much more
likely.

A top-level fault tree analysis for explosion and fire (Figure 8-1) shows
many causes that can be easily controlled. Others will require redesigns.

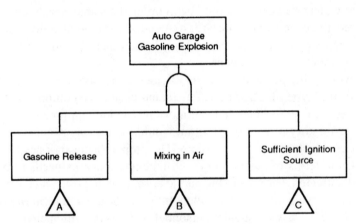

FIGURE 8-1 Auto Garage Gasoline Explosion FTA. (*Figure continues on pp. 209–211*)

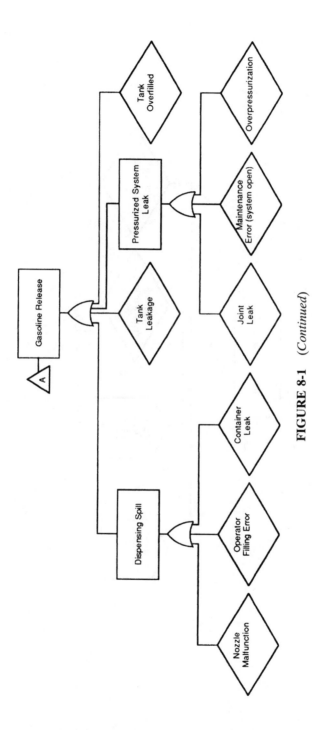

FIGURE 8-1 (*Continued*)

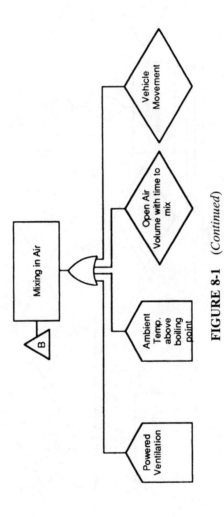

FIGURE 8-1 (*Continued*)

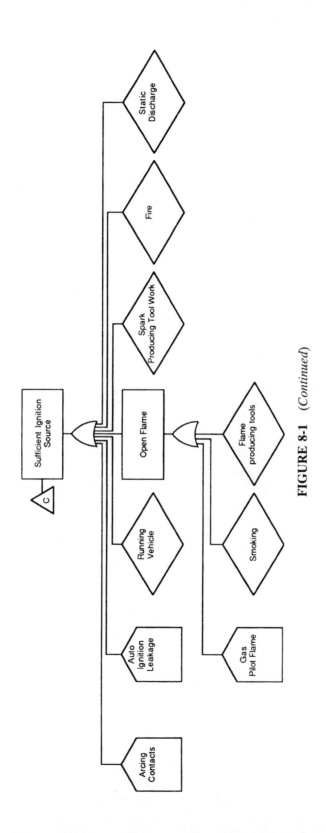

FIGURE 8-1 *(Continued)*

REFERENCES

Abercrombie, Stanley A., 1983. Enlarging the Focus on Motor Fleet Safety. *Professional Safety*, July: 15–19.

Abercrombie, Stanley A., 1980. Readers' Pulse—Buckle Up–Every Time! *Professional Safety*, April: 10.

Adams, Carl E., Jr., 1991. Industrial Wastewater Treatment Options. *National Environmental Journal*, Sept./Oct.: 16–18.

Adams, J. Robert, 1976. *Risk Control 3* (Supplement 29): 135–137.

Allison, William W., 1991. Other Voices: Are We Doing Enough? *Professional Safety*, Feb.: 31–32.

American Industrial Hygiene Journal, 1991, Feb.: A-92–94. Labor Report: BLS Reports on Survey of Occupational Injuries and Illnesses in 1989.

American Red Cross, 1987. *Adult CPR Workbook*.

Anderson, Charles K. & Catterall, Mary J., 1987. A Simple Redesign Strategy for Storage of Heavy Objects. *Professional Safety*, Nov.: 35–38.

Anna, Daniel H., 1989. Exposure to Low Frequency Electromagnetic Fields and Cancer Development. *Professional Safety*, Nov.: 40–42.

Askren, William B. & Howard, John M., 1988. Software Safety Lessons Learned from Computer-Aided Industrial Machine Accidents. *Hazard Prevention*, May/June: 28–30.

Ayoub, M. M., Jiang, B. C., Selan, J. L., & McDaniel, J. W., 1987. Establishing a Physical Criterion for Assigning Personnel to U.S. Air Force Jobs. *American Industrial Hygiene Association Journal*, May: 464–470.

Backastow, Jack L., 1978. *Accident/Incident Investigation Techniques*, LA-7877-MS, Oct.: Appendix III.

Banks, W. W. & Cerven, F. 1985. Predictor Displays: The Application of Human Engineering in Process Control Systems. *Hazard Prevention*, Jan./Feb.: 26–32.

Benner, Ludwig, Jr., 1990. Safety Training Achilles' Heel. *Hazard Prevention,* 1: 6–12.

Benner, Ludwig, Jr., 1982. 5 Accident Perceptions—Their Implications for Accident Investigators. *Professional Safety,* Feb.: 21–27.

Benson, John D., 1987. Control of Low Back Pain Using Ergonomic Task Redesign Techniques. *Professional Safety,* Sept.: 21–25.

Bird, Frank E., Jr., (Ed.), 1973. *Selected Readings in Safety.* Macon, GA: Academy Press.

Bonner, Robert E., 1987. Driver Records Predict Accidents. *Professional Safety,* Nov.: 32–34.

Bottom Line Personal, 1989. Mar. 15: 9.

Bukowski, Richard W., 1985. Researchers Providing Some Details About Hazards of Fire. *Chemical & Engineering News,* 20 May: 35–38.

Bullock, Collin, Mitchell, Frank, & Skelton, Bob, 1991. Development in the Use of the Hazard and Operability Study Technique. *Professional Safety,* Aug.: 33–40.

Bureau of National Affairs, 1989. *Occupational Safety & Health: 7 Critical Issues for the 1990s* (BSP-136), July.

Burks, Dennis, 1990. Video Tools of the Trade. *Occupational Hazards,* Nov.: 53–54.

Burton, D. Jeff, 1986. Cooperative Program Develops Engineering Controls for Lead Industry. *Occupational Health & Safety,* Sept.: 48–49.

Campbell, Chris W., 1990. All About Confined Spaces. *Professional Safety,* Feb.: 33–35.

Carr, T. S., 1991. Underground Mine Disasters—History, Operations and Prevention. *Professional Safety,* March: 28–32.

Chaffin, Don B., 1987. Manual Materials Handling and the Biomechanical Basis for Prevention of Low-Back Pain in Industry—An Overview. *American Industrial Hygiene Association Journal,* Dec.: 989–996.

Chang, Insun, 1986. Laser Safety. *Professional Safety,* Nov.: 50–53.

Chapnik, Elissa-Beth & Gross, Clifford M., 1987. Evaluation, Office Improvements Can Reduce VDT Operator Problems. *Occupational Health & Safety,* July: 34–37.

Clarke, Arthur C., 1968. *2001: A Space Odyssey.* New York: Signet.

Clarke, Arthur C., 1982. *2010: Odyssey Two.* New York: Ballantine.

Clark, Carl C., 1984. *A Review of Crash Protection of Children, the Elderly, and the Handicapped, and the Design of Inflatable Compartmentalized Controlled Deformation Protective Systems for Them.* Presented at the Third International Conference on Mobility and Transport for Elderly and Handicapped Persons. Orlando, FL, October.

Colonna, Guy R., 1987. Confined Space Hazards Dictate Employee Training Improvements. *Occupational Health & Safety,* July: 21–28.

Colver, C. Phillip & Colver, John C., 1991. Managers, Workers Must Realize Need for Flame-Retardant Clothing. *Occupational Health & Safety,* Jan.: 20–23.

Coudal, Edgar F., 1988. Power Line Protection: When, Why and What Kind? *Engineer's Digest,* May: 10–12.

Cross, Rich, 1990. The Feds' Master Plan for Highway Safety. *Professional Safety,* Oct.: 38–42.

Cunitz, Robert J. & Middendorf, Loarna, 1985. Problems in the Perception of Overhead Power Lines. *Hazard Prevention,* March/April.

Damon, Mark S., 1981. *Techniques of Fall Protection.* Paper presented at the ASSE Region III Professional Conference, Texas A&M University, February.

Darzinskis, Kaz, 1990. Readers' Pulse. *Professional Safety,* Feb.: 12.

Davé, Niru, 1989. Safety Policy Statement: Developing and Using It, *Professional Safety,* June: 29–32.

DB Industries, Inc., 1983. *Impact Test of Lanyards.* St. Paul, MN, March 15.

Della-Giustina & Deay, Ardeth, 1991. New Developments for Safety Training Programs Include Instructional Enhancements. *Professional Safety,* Jan.: 17–20.

Densmore, Bill, 1982. Kieland Victims, Families Sue Phillips for $2 Billion. *Business Insurance,* March 15: 3, 33.

Dept. of Health and Human Services. *Guidelines for Controlling Hazardous Energy During Maintenance and Servicing.* NIOSH Publication No. 83-125.

Dept. of Health and Human Services, 1983. *NIOSH Current Intelligence Bulletin 38.* NIOSH No. 83-110, Mar. 29.

Dept. of Labor, 1989. *OSHA/State Consultation Project Directory.*

Dept. of Transportation, 19: 18. Sudden Impact—*An Occupant Protection Fact Book.* National Highway Safety Administration.

DiMaggio, Angelo & Mooney, Vert, 1987. Conservative Care for Low Back Pain: What Works? *Journal of Musculoskeletal Medicine,* Sept.: 27–31.

Dittman, Charles R., 1988. Selling Management the Best Fire Protection Available. *Professional Safety,* Oct.: 21–26.

Eisma, Teri Lyn, 1990. Loading Dock Safety Begins with Design. *Occupational Health & Safety,* Feb.: 49–51.

Eisma, Teri Lyn, 1990. Rules Change, Worker Training Helps Simplify Fall Prevention. *Occupational Health & Safety,* Mar.: 52–54.

Eisma, Teri Lyn, 1990. Demand for Trained Professionals Increases Educational Interests. *Occupational Health & Safety,* Stevens Publishing, Waco, TX: 52–56.

Ellis, J. Nigel, 1986. *How to Develop an Effective Fall Protection Program.* Vallen Forum.

Emerson, Brian, 1985. The Accident Phenomenon. *Professional Safety,* April: 20–26.

Everett, Bruce E., 1989. Training Techniques That Work Within an Integrated Safety Program. *Professional Safety,* May: 34–37.

Family Safety, 1982 Summer.

Fawcett, Howard K. & Wood, William S. (Eds.), 1982. *Safety and Accident Prevention in Chemical Operations.* New York: John Wiley & Sons.

Ferry, Ted S., 1981. Accident Investigation and Analysis. *Professional Safety,* Jan.: 18–22.

Folkers, Karl, 1986. Contemporary Therapy with Vitamin B_6, Vitamin B_2, and Coenzyme Q_{10} (Priestly Medal Address). *Chemical & Engineering News,* 21 April: 27–56.

Forsberg, Krister & Farriadis, Stathoula, 1986. The Permeation of Multi-Component Liquids Through New and Pre-Exposed Glove Materials. *American Industrial Hygiene Association Journal,* May: 189–193.

Franklin, Frederick F., 1990. Circuit Breakers: The Myth of Safety. *Professional Safety,* June: 28–31.

Freeman, Raymond A., 1992. Documentation of Hazard and Operability Studies. *Hazard Prevention,* 1: 14–17.

Fulwiler, Richard D. & Hackman, Richard J., 1990. EPA's Intervention into Workplace Health and Safety—The Other OSHA. *American Industrial Hygiene Association Journal,* July: A-490–A-495.

Goel, Vijay K. & Rim, Kwan, 1987. Role of Gloves in Reducing Vibration: An Analysis for Pneumatic Chipping Hammer. *American Industrial Hygiene Association Journal,* Jan.: 9–14.

Goyal, Ram K. & Al-Jurashi, Nassir M., 1990. Gas Dispersion Models. *Professional Safety,* May: 23–33.

Greenberg, Leo, 1979. The Safe Design of Production Systems. *Hazard Prevention,* Sept./Oct.: 24–25.

Gressel, Michael G. & Gideon, James A., 1991. An Overview of Process Hazard Evaluation Techniques. *American Industrial Hygiene Association Journal,* April: 158–163.

Groneman, Louise, 1985. Carpal Tunnel Syndrome Can Be Lessened with Early Treatment. *Occupational Health & Safety,* Oct.: 39–46.

Gruber, George J., 1976. Relationships between Wholebody Vibration and Morbidity Patterns among Interstate Truck Drivers. NIOSH publication No. 77-167, November.

Hair, Dan M., 1991. Farm Safety. *Professional Safety,* Oct.: 17–22.

Hall, Ed, 1987. Protective Eyewear, Proper Care Can Help Stop Injuries, Blindness at Work. *Occupational Health & Safety,* Jan.: 70–80.

Hall, Stephen K., 1988. Industrial Chemical Disasters. *Professional Safety,* July: 9–13.

Hardie, Douglas T., 1981. The Safety Professional of the Future. *Professional Safety,* May: 17–21.

Hazard Prevention, 1987 Jan./Feb.: 29–32. A Survey of Electrical Fires.

Heyl, Frank, 1986. Protection When the Mercury Plunges. *Occupational Health & Safety,* Dec.: 44–48.

Hight, Sterling K., 1981. Assessment of Product Liability Exposure. *Professional Safety,* July: 22–26.

Hislop, Richard D., 1991. A Construction Safety Program. *Professional Safety,* Sept.: 15–20.

Horner, Jack, 1982. You Can Mow Safely Now. *Family Safety,* Summer: 26–28.

Hoyt, William R., 1984. Carpal Tunnel Syndrome: Analysis and Prevention. *Professional Safety,* Nov.: 16–21.

Hurst, Ron & Khalil, Tarek, 1984. Entering and Exiting Elevated Vehicles. *Professional Safety,* Sept.: 20–26.

Industrial Health & Safety News, 1988 Aug: 14. News: Employee Wins VDT Settlement.

Industrial Hygiene News, 1992 Sept.: 48. Attention to Fluid Replenishment Can Reduce Heat Stress Risks.

Industrial Safety & Health News, 1990: 23. Washington Outlook: OSHA vs. EPA? (White Paper).

Industrial Safety & Hygiene News, 1984 July: 17. Integrating PPE Systems.

Jenkins, Larry, 1984. *Combustion Toxicology.* Presented to the Houston Industrial Hygiene Council, August 16.

Job Health Highlights—3M, 1991 Summer: 7. EPA AND OSHA Agree on Joint Inspections.

Johnson, Kim, 1985. Analytical Report on the Causes and Prevention of Carpal Tunnel Syndrome. *Professional Safety,* Oct.: 48–51.

Johnson, Robert D., 1984. Accident Investigation. *National Safety News,* Sept.: 70–74.

Johnson, William G., 1980. *MORT Safety Assurance Systems.* New York: Marcel Dekker.

Kam, John Dai-Hong, 1981. Effects of Whole-Body Vertical Vibration on Manual Dexterity Test Results. *American Industrial Hygiene Association Journal,* July: 556–557.

Kappler, George P., Jr., 1988. Limited Use Garments Have Expanded Role in Protective Clothing Industry. Occupational Health & Safety, Aug.: 35–36.

Kasper, Ronald J., 1982. Minimizing Potential Damage from Cable Tray Fires. *Plant Engineering,* 18 Feb.: 91–94.

Keyserling, W. Monroe, Herrin, Gary D., Chaffin, Don, Armstrong, Thomas J. & Foss, Merle L., 1980. Establishing an Industrial-Strength Testing Program. *American Industrial Hygiene Association Journal,* Oct.: 730–736.

Khoudja, Dadir-Ben-Ali & Johnson, Waymon L., 1986. A Computerized Analysis Technique for Evaluating Fire Safety of Buildings. *Hazard Prevention,* March/April: 9–14.

Kleronomos, C. C., 1976. Electrical Hazard Detection and *OSHAct* Compliance. *National Safety News,* Sept.: 68–72.

Krause, D. L., 1985. Forensic Investigations. *Professional Safety,* April: 28–31.

Krikorian, Michael, 1978. Implementing Hiring the Handicapped. *National Safety News,* Sept.: 87–89.

Kroemer, K. H. E., 1992. Avoiding Cumulative Trauma Disorders in Shops and Offices. *American Industrial Hygiene Association Journal,* Sept.: 596–604.

Kubias, F. Owens, 1989. Reader's Pulse: Engineering Courses. *Professional Safety,* Oct.: 10.

Kuhlman, Raymond, 1986. The Need for Off-the-Job Safety Programs. *Professional Safety,* Mar.: 13–16.

Kuhlman, Raymond L., 1988. Eight Degrees from Death: Notes and Comments. *Professional Safety,* Nov.: 33.

Kuhlman, Raymond L., 1989. Notes and Comments: Get Off on the Right Foot. *Professional Safety,* March: 41.

La Bar, Gregg, 1990. Electromagnetic Fields: The Problem with Power. *Occupational Hazards,* Oct.: 91–95.

La Bar, Gregg, 1990. How to Improve Your Accident Investigations. *Occupational Hazards,* Mar.: 35.

La Bar, Gregg, 1990. OSHA Plans Update of PPE Rules. *Occupational Hazards,* June: 51–53.

La Bar, Gregg, 1991. Getting Work Off Employees' Backs. *Occupational Hazards,* April: 27–29.

La Bar, Gregg, 1991. OSHA's Mission Expands. *Occupational Hazards,* Sept.: 95–97.

La Bar, Gregg, 1991. Spreading the Net. *Occupational Hazards,* Sept.: 99–102.

La Bar, Gregg, 1991. Worker Training: An Investment in Safety. *Occupational Hazards,* Aug.: 23–26.

Lahey, James W., 1983. Nets: Watch That First Step. *National Safety News,* Nov.: 70.

Lahey, James W., 1988. Hands: The $5.5 Billion Challenge. *Safety & Health,* June: 36–38.

Lawn, George W., 1987. Times and Treatments Change in Back Care. *Professional Safety,* April: 28–32.

Leveson, Nancy G., 1986. Software Safety: *Why, What and How* (Technical Report 86-04). Information and Computer Science Dept., University of California, Irvine, February.

Liles, D. H. & Mahajan, P., 1985. Using NIOSH Lifting Guide Decreases Risks of Back Injuries. *Occupational Health & Safety,* Feb.

Lingenfelter, Gerald E., 1988. Evaluating Product Safety Certification Programs. *Professional Safety,* Feb.: 13–18.

McDermott, Robert F., 1985. Airbags and Seatbelts—Why Not Both? *USAA Aide Magazine,* Summer: 10–12.

McCormick, Ernest J. (Ed.), 1970. *Human Factors Engineering,* 3rd ed., p. 3. New York: McGraw-Hill.

McDonald, James M., 1989. A Short Guide to Choosing Level A Protective Clothing. *Industrial Hygiene News,* Sept.: 54–55.

McElroy, Frank E. (Ed.), 1980. *NSC Accident Prevention Manual for Industrial Operations: Engineering and Technology,* 8th ed. International Fire Chiefs Association, Merrifield, Va.

McKinney, C. Don, 1983. A Solution to Juvenile Fire Setting in Houston. *International Fire Chief,* Aug.: 78–80.

Man, 1985 Dec.: 22–28. Indisputably Viable, Increasingly Affordable: Laser Processing Is Here to Stay.

Man, Vincent A., Bastecki, Vincent, Vandal, Grace, & Bentz, Alan P., 1987. Permeation of Protective Clothing Materials: Comparison of Liquid Contact, Liquid Splashes and Vapors on Breakthrough Times. *American Industrial Hygiene Association Journal,* June: 551–555.

Mattheis, Darrell K., 1992. A National MSDS Repository. *American Industrial Hygiene Association Journal,* Sept.: 607–608.

Merriman, Melaney, 1990. Equipment Designed for Right Handers Leaves Safety Doubts for Southpaws. *Occupational Health & Safety,* Aug.: 25–26.

Meshkati, Najmedin, 1990. Preventing Accidents at Oil and Chemical Plants. *Professional Safety,* Nov.: 15–18.

Metry, Amir A. & Fazzini, Alexander J., 1980. Stabilizing Hazardous Industrial Wastes. *Plant Engineering,* 4 Sept.: 85–87.

Miller, Barret C., 1989. Biological Effects of Exposure to Electromagnetic Fields. *Professional Safety,* Aug.: 21–23.

Minter, Stephen G., 1990. A New Perspective on Head Protection. *Occupational Hazards,* June: 45–49.

Minter, Stephen G., 1990. Out of Control? (Editorial). *Occupational Hazards,* July: 7.

Minter, Stephen G., 1990. Sizing Up Safety (Editorial). *Occupational Hazards,* Sept.: 7.

Mintz, Benjamin W., 1989. Disorder and Early Sorrow in the OSHA program. *American Industrial Hygiene Association Journal,* Feb.: A-96–A-108.

Morin, Catherine J., Cleaver, Robert B., & Batinger, William F., 1988. Safety Officers Beginning to Require the Use of Treated Work Garments. *Occupational Health & Safety,* Aug: 26–33.

NASA, JSC, 1989. *Reliability/Problem Trend Analysis and Reporting System (RPTARS), NSTS 08341, Vol. III* (S41643 Attachment), April 14.

Nassif, George B., 1991. Products Liability and the European Economic Community. *Professional Safety,* Jan.: 21–23.

National Safety Council, 1989. *Accident Facts, 1989 Edition.*

National Safety News, 1980 July: 41–44. What You Should Know About Electrical Safety.

National Safety News, 1981 Jan.: 30–32. Talk About Fire: Hotel Fire Safety.

National Safety News, 1983 Aug: 61–69. Safety Nets: Fall Protection for the Construction Industry. National Safety Council Data Sheet I-608-Rev. 83.

National Safety News, 1988 June: 55–58. Selection and Use Guidelines for Chemically Impervious Gloves. NSC Data Sheet 1-735-87.

Nemec, Margaret M., 1990. What Can You Do About Back Injuries? *Occupational Hazards,* Mar.: 62–66.

Nertney, R. J. & Buys, J. R. 1976. Training as Related to Behavioral Change. *ERDA-76-45-6,* June.

Network, 1982 Summer. Water Emergencies!

Novitski, David G. & Trattner, Richard B., 1991. Groundwater Treatment Choices. *National Environmental Journal,* Sept./Oct.: 38–41.

Occupational Hazards, 1989 Dec.: 22. OSHA Communique: OSHA Proposes Highest Penalties Ever Against USX Corp.

Occupational Hazards, 1989 Dec.: 29. Scannell Charts OSHA's Future Course.

Occupational Hazards, 1990 July: 13. EPA—Electromagnetism and Cancer May Be Linked.

Occupational Hazards, 1990 Sept.: 11. OSHA's Drive for Motor Vehicle Safety.

Occupational Hazards, 1990 Dec.: 25. Safety & Health: Worker Fatalities on the Decline.

Occupational Hazards, 1991 June. Ergonomics Standards a Long Way Off.

Occupational Hazards, 1991 June: 11. Egregious Policy: An Obsolete Weapon?

Occupational Hazards, 1991 June: 12–13. NIOSH Lifting Guide Updated.

Occupational Hazards, 1991 Sept.: 192. Contractor Safety Program: 10 Elements for Success (DuPont advertisement).

Occupational Health & Safety, 1978 Mar./April: 8. News: Health and Safety Staffs to Triple?

Occupational Health & Safety, 1985 June: 5–6. News: Cancer, Back Pain Lead NIOSH List.

Occupational Health & Safety, 1989 Dec.: 10. News: Standards—OSHA Revises Trenching Rule.

Occupational Health & Safety, 1990 July: 9. News: APTA Offers Advice on Ergonomics.

Occupational Health & Safety, 1990 Dec.: 9–10. News: OSHA Fines Will Soar Under New Budget.

Occupational Health & Safety, 1991 Jan.: 40. News: Report Analyzes Construction Deaths.

O'Neill, Brian, 1988. Auto Safety—Seat Belt Most Effective When Worn Correctly. *USAA Aide Magazine,* April: 8.

Orzech, Mary A., Goodwin, Mark D., Brinkley, James W., Salerno, Mark D., & Seaworth, John, 1987. *Test Program to Evaluate Human Response to Prolonged Motionless Suspension in Three Types of Fall Protection Harnesses.* Air Force Systems Command, AAMRL-TR-055, September.

OSHA, 1987. *Excavation Work Proposal,* April 15.

Ostrom, Lee T., Hill, Susan G., Close, James A., Lash, James D., & Hemming, William C., 1991. Developing an Emergency Procedures Writers' Guide: A Case Study. *Proceedings of the Tenth International System Safety Conference,* Vol. 1, pp. 2-1–3-15. Dallas, July 18–22.

Owen, Bernice D., 1986. Posture, Exercise Can Help Prevent Low Back Injuries. *Occupational Health & Safety,* June: 33–37.

Pagnelli, David, 1990. Carpal Tunnel Syndrome: The Multi-Industry Health Hazard. *Industrial Hygiene News,* Jan.: 38–39.

Pater, Robert F., 1988. How to Make High Impact Safety Presentations. *Professional Safety,* Feb.: 19–22.

Pater, Robert, 1985. Fallsafe: Reducing Injuries from Slips and Falls. *Professional Safety,* Oct.: 15–18.

Paques, J. J., Masse, S., & Belanger, Raymond, 1989. Accidents Related to Lockout in Quebec Sawmills. *Professional Safety,* Sept.: 17–20.

Pearson, George W., 1989. Robotics: An International Health and Safety Solution. *Professional Safety,* Dec.: 28–31.

Pearson, Richard E., 1987. Applying the System Safety Concept to Control Confined Space Entry and Hazards. *Professional Safety,* May: 28–30.

Peterson, Dan, 1981. Role and Function of the Safety Manager. *Occupational Health & Safety,* Sept.: 63–66.

Pettit, Ted A., Sanderson, Lee M., and Linn, Herbert I., 1987. Workers/Rescuers Continue to Die in Confined Spaces. *Professional Safety,* Feb.: 15–22.

Pinyan, Dugold, 1988. Emerging Trends. *Safety & Health,* April: 38–40.

Pliakos, Mark, 1992. Software Safety and Systems Safety. *Hazard Prevention,* 3: 14–19.

Professional Safety, 1985 April: 11–12. Safety Digest: VDTs Are Not Harmful to Pregnant Women.

Professional Safety, 1987 Dec.: 1. Upfront: Driving More Than Eight Hours Is Dangerous for Truck Drivers.

Professional Safety, 1988 Feb.: 1. Upfront: FWHA Proposes Uniform Testing Standards for Truck and Bus Drivers.

Professional Safety, 1988 Apr: 8. OSHA News.

Professional Safety, 1988 Aug.: 1. Upfront: From '70 to '86, DUI Arrests Jumped 127%.

Professional Safety, 1989 Feb.: 8. OSHA News: OSHA Studies Workplace Deaths Involving Welding.

Professional Safety, 1989 March: 10. Safety Digest: Highways Designed for Cars—Not Big Rigs.

Professional Safety, 1989 Aug.: 30. Safety Digest: Rear Seat Lap/Shoulder Safety Belts.

Professional Safety, 1989 Oct.: 12. Safety Digest: Highway Safety Leadership Council.

Professional Safety, 1989 Oct.: 12. Safety Digest: Stiffer Nationwide Drunk Driving Enforcement.

Professional Safety, 1989 Dec.: 12. Safety Digest: Wheel Retainers.

Professional Safety, 1990 Feb.: 1. Upfront: Workplace Health, Safety Topics to Watch in '90s.

Professional Safety, 1991 Jan.: 1, 39. Upfront: Age Not a Factor in Frequency of Back Injuries, Researchers Say.

Professional Safety, 1991 Jan.: 39. Upfront: Arson Fires, Related Deaths and Losses Down in 1989.

Professional Safety, 1991 Feb.: 12. Safety Digest: Decade of Progress in Motor Vehicle Safety.

Professional Safety, 1991 May: 1. Upfront: Laser-related Injuries a Growing Concern.

Radwin, Robert G., Armstrong, Thomas J., & Vanberqeuk, Ernst, 1990. Vibration Exposure for Selected Power Hand Tools Used in Automobile Assembly. *American Industrial Hygiene Association Journal,* Sept.: 510–518.

Rasmussen, Jens, 1988. Approaches to the Control of the Effects of Human Error on Chemical Plant Safety. *Professional Safety,* Dec.: 23–29.

Rekus, John F., 1990. Invisible Confined-Space Hazards Require Comprehensive Entry Program. *Occupational Health & Safety,* Aug.: 38–51.

Rekus, John F., 1990. OSHA's Lockout-Tagout Standard Mandates Control of Energy Sources. *Occupational Health & Safety,* Oct.: 108–119.

Ring, Leonard, 1989. Back Injury Prevention—Awareness versus Performance. *Professional Safety,* July: 11–14.

Rodriguez, Albert P., 1980. How 'Special Disability' Law Impacts Hiring Handicapped. *Risk Management,* Sept.: 50–51.

Rosenwasser, Teri L., Potter, John W., & Parr, Richard B. Vision Losses Prevented by Using Protective Eyewear. *Occupational Health & Safety,* April: 63–66.

Rowe, M. Laurens, 1982. Are Routine Spine Films on Workers in Industry Cost- or Risk-Benefit Effective? *Journal of Occupational Medicine,* Jan.: 41–43.

Ruskouski, Charles, 1990. Emergency Lighting Equipment: A Part of Every Life Safety Program. *Professional Safety,* June: 32–34.

Ryan, Joseph P., 1988. Hazard Analysis Guidelines in Product Design. *Professional Safety,* Mar.: 17–19.

Safety & Health, 1988 Apr.: 61. Safety Clips: Spinal Injuries.

Safety & Health, 1988 June: 23. Firenews: Causes of Fire Deaths.

Samways, Margaret C., 1987. OSHA Voluntary Guidelines Provide Blueprint for Employee Training. *Occupational Health & Safety,* May: 68–74.

Sandman, Peter M., 1990. *Risk Communication.* Paper presented at the American Industrial Hygiene Convention and Exposition. Orlando, Florida, May 14.

Sawyer, Kenny, 1987. An On-site Exercise Program to Prevent Carpal Tunnel Syndrome. *Professional Safety,* May: 17–20.

Scheel, Paul D., 1993. Robotics in Industry. *Professional Safety,* Mar.: 28–32.

Schulz, Bill, 1991. Deer Stands Hazardous to Hunters. *The Huntsville Times,* Sept. 8, p. D14.

Sheridan, Peter J., 1990. State Plans Offer Enforcement Choices. *Occupational Hazards,* June: 70–73.

Sheridan, Peter J., 1991. How to Handle an OSHA Inspection. *Occupational Hazards,* Sept.: 132–133.

Sheridan, Peter J., 1991. Is OSHA Due for an Overhaul? *Occupational Hazards,* Sept.: 85–89.

Sheridan, Peter J., 1991. Meatpackers Move to Cut Injury Rates. *Occupational Hazards,* May: 81–85.

Sheridan, Peter J., 1991. OSHA Calling. *Occupational Hazards,* Sept.: 129–133.

Sliney, David H., 1986. Laser Safety: The Newest Face on an Old Standard. *Photonics Spectra,* April: 83–96.

Smith, Gary D., 1991. A Standard of Care Extended to Third-Party Non-direct Employers on Multiple Employer Work Sites. *Professional Safety,* July: 35–44.

Smith, R. Blake, 1990. Manufacturing Companies of All Sizes Benefit from Waste-Reduction Policy. *Occupational Health & Safety,* Nov.: 19–21.

Smith, R. Blake, 1991. When Ergonomics Was Young. *Occupational Health & Safety,* Jan.: 1.

Smith, S. L., 1990. Steel-Toed Safety. *Occupational Hazards,* Nov.: 48–51.

Snook, Stover H., 1980. The Design of Manual Handling Tasks. *Professional Safety,* May: 18–26.

Snook, Stover H., 1988. Approaches to the Control of Back Pain in Industry: Job Design, Job Placement, and Education/Training. *Professional Safety,* Aug.: 23–31.

Snook, Stover H., Campanelli, Ralph A., & Hart, Joseph W., 1978. Three Preventive Approaches to Low Back Injury. *Professional Safety,* July: 34–38.

Sorock, Gary, 1981. *A Review of Back Injury Prevention and Rehabilitation Research: Suggestions for New Programs* (for National Safety Council Back Injury Committee), May: 2.

Stanevich, Ronald L. & Middleton, Dannie C., 1988. An Exploratory Analysis of Excavation Cave-in Fatalities. *Professional Safety,* Feb.: 24–28.

Starck, Jukka, Jussi, Pekarinen, & Ilmari, Pyykko, 1990. Physical Characteristics of

Vibration in Relation to Vibration-Induced White Finger. *American Industrial Hygiene Association Journal,* April: 179–184.

Stevens, Arthur M., 1985. Routine Inspections, Maintenance Can Prevent Industrial Fire Mishaps. *Occupational Health & Safety,* Dec.: 12–14.

Sulowski, Andrew C., 1979. Selecting Fall Arresting Systems. *National Safety News,* Oct.: 55–60.

Suruda, Anthony, 1988. Electrocution at Work. *Professional Safety,* July: 27–32.

Talty, John T. & Walters, James B., 1987. Integration of Safety and Health into Business and Engineering School Criteria. *Professional Safety,* Sept.: 26.

Tanner, Melissa, 1990. Increasing Use, Power of Lasers Make Eye Protection Essential. *Occupational Health & Safety,* July: 44–46.

Traffic Safety, 1982 Nov./Dec.: 31. Drunk Driving: The Harsh Realities and Practical Solutions—Drinking Age.

Travelers Product Liability Newsletter, 1978 Oct.: 2–3.

Tyson, Patrick R., 1988. The Official Coin Toss. *Safety & Health,* April: 57–58.

USAA Aide Magazine, 1985 Fall: 31. Safety Tips: Fireplace Safety.

USAA Aide Magazine, 1986 Fall: 8. Helmet Use: Some Statistics to Think About.

USAA Aide Magazine, 1989, June: 18–21. Off-the-Job Safety (Accidents and Children).

USAA Aide Magazine, 1989 Oct.: 13. Safety—Avoid Accidents through Safe Driving Habits.

USAA Aide Magazine, 1989 Oct.: 25. Health & Safety: The Importance of Wearing Helmets.

USAA Aide Magazine, 1990 April: 17–19. Safety—Survivor or Statistic? How 103 Automobiles Compare.

USAA Aide Magazine, 1990 Aug.: 32. Health & Safety: Senior Safe.

USAA Aide Magazine, 1990 Oct.: 24. Health & Safety—The Safest Color?

USDL News Release 88-223, 1988 April 29.

Vahdat, Nader & Delaney, Reginald, 1989. Decontamination of Chemical Protective Clothing. *American Industrial Hygiene Association Journal,* March: 152–156.

Vincoli, Jeffrey W., 1988. OSHA Investigation Refresher. *Professional Safety,* Mar.: 24–29.

Walsh, M. L., Harvey, S. M., Facey, R. A., & Mallette, R. R., 1991. Hazard Assessment of Video Display Units. *American Industrial Hygiene Association Journal,* Aug.: 324–331.

Weaver, L. A., III, 1979. Vibration: An Overview of Documented Effects on Humans. *Professional Safety,* April: 29–37.

Weiss, Michael J., 1988. Safety—The Winning Combination. *USAA Aide Magazine,* Feb.: 6–7.

Westchurchman, C., 1968. *The Systems Approach,* p. 35. New York: Dell.

White, David, 1987. Boilovers. *Industrial Fire World,* Feb.: 8–12.

Wilcher, Frank W., Jr., 1987. Performance Certification of Clothing Connected to Concern for Protection. *Occupational Health & Safety,* June: 39–44.

Yokel, Felix Y. & Ching, Riley M., 1983. Proposed Standards for Construction Practice in Excavation. *Professional Safety,* Sept.: 34–39.

Yoxall, Patricia, 1982. Driving Can Be Risky. *Family Safety,* Summer: 8–10.

INDEX

Accident, defined, 10
Accident investigation:
  barriers, questions regarding, 24
  change analysis, 24–25
  final documentation, 26–27
  four p's of, 23
  function of, 20–21
  fundamental questions, 24–26
  personnel and methodology, 22–24
  plans, 21–22
  twelve-step process, 23
Accident-investigation manual, 22
Action, problem recognition *vs.*, 13
Agricultural safety:
  controls, 29–30
  hazards, 28–29
  statistics, 27–28
Air Force Specialty Codes (AFSCs), 33
American Board of Industrial Hygiene
  (ABIH), 4
American Conference of Governmental
  Hygienists (ACGIH):
  noise limitation standards, 92
  Threshold Limit Values, 80
American Industrial Hygiene
  Association (AIHA), 4
American Institute of Chemical
  Engineers (AIChE), Center for
  Chemical Process Safety, 39–40
American National Standards Institute
  (ANSI):
  confined space standards, 41
  fall protection equipment, 72–73

  laser safety standards, 97
American Society for Testing and
  Materials (ASTM), 41
American Society of Safety Engineers
  (ASSE), 4
Americans with Disabilities Act (ADA),
  68
Assured Equipment Grounding
  Conductor assurance program,
  52

Back injuries:
  causes, 31
  controls, 33–34
  effects, 30–31
  research, lifting and ergonomics,
   31–33
  task redesign principles, 35–37
  treatment, 34–35
  types of, 29
Back Schools, 34
Bleeding, first aid procedures, 60
Board of Certified Safety Professionals
  (BCSP), 4
Boiling liquid expanding vapor
  explosion (BLEVE), 38
Building codes, fire safety and, 77
Bureau of National Affairs, Inc.
  (BNA), 121–122
Burns, first aid procedures, 60
Business Round Table, Construction
  Industry Cost Effectiveness Project,
  42, 49

Carpal Tunnel Syndrome (CTS), 66–67
Cars:
  industrial hygiene and, 113
  transportation safety, and 167
Catastrophes:
  cause and effects, 39
  control, 39–40
  defined, 37
  occurrences, 38
  response, 38–39
Chain-of-events perception, 20
Change management (CM), 190–191
Chemical Abstract Service (CAS), 172
Chemical operations, see Hazardous
    operations (HazOps)
Children, injury sources, 109
Class I lasers, 97
Class II lasers, 97
Class IIIB lasers, 98
Class IV lasers, 98
Clean Air Act, 119, 172
Clean Water Act, 172
Communication, significance of,
    106–107, 177
Computer technology, impact of, 64, 149
Conceptual preliminary hazard
    analysis (CPHA), 157–159, 163
Cone of Learning, 143–144
Confined space safety:
  agriculture industry, 28
  controls, 41
  documentation, 41–42
  hazards, 40–41
Construction safety:
  construction lasers, 47
  contractor safety, 47–49
  contracts, 47
  electrical hazards, 45
  elevated work, 45–46
  fire hazards, 46
  materials handling safety, 46–47
  trenching/excavation, 42–45
Consumer Product Safety Act (CPSA),
    133
Consumer Product Safety Commission
    (CPSC), 134
Contractor safety, 47–49
Cooperative Assessment Program
    (CAP), 120

Corrected Effective Temperature (CET),
    90
CPR, as emergency response, 61
Critical Incident Techniques, 16
Cryogenic safety:
  controls, 50–51
  hazards, 50, 131
Consumer Product Safety Act (CSPA),
    133
Cumulative trauma disorders:
  causes, 67
  controls, 67–68
  mining safety and, 108
  overview, 66–67

Deductive reasoning, 155–156
Determinant variable, 20
Dilution ventilation, 108
Disabled workers, 68
Documentation, see Safety
    documentation
  accident investigation, 26–27
  confined space safety, 41–42
  lockout/tagout records, 105
Dosimeters, 92–93
du Pont de Nemours, safety program, 14

Electrical Power Research Institute
    (EPRI), 53
Electrical safety:
  construction work, 45
  controls, 52–53
  electrically-related safety effects,
    53–54
  electromagnetic fields, 53
  hazards, 51–52
  welding and, 174
Electromagnetic interference (EMI), 53
Emergency response (ER):
  defined, 54
  emergency, phases of, 54–55
  first aid, 57–62
  industrial hygiene and, 113
  office safety, 114
  plan development, 55–57
  relationships, 55
Energy exchange, 15
Engineering applications, case studies:
  banking industry, 201–203, 206–207

golf course maintenance shop, 202–205
parking garage, 205, 207
Environmental Protection Agency (EPA), 53, 81, 119, 171–172
Ergonomics:
  agricultural safety, 31–33, 35–37
  applications:
    cumulative trauma disorders, 65–67
    ergonomic design, 65
    materials handling, 68
    VDT/VDU effects, 65–66
  causes, 67
  control system, 63–65
  defined, 62–63
  mining safety, 108
E-teams (emergency teams), 117
Evaluation, timing of, 15–16
Eye protection, equipment for, 51

Face shields, 51
Facility safety, 69–71
Factorial perspective, 20
Fall Arresting System, 72
Fall protection:
  agriculture industry and, 30
  construction work and, 45–46
  controls, 73–74
  falls from elevations, 71–72
  falls from same elevation, 72
  systems, active and passive, 72–73
Farm machinery and equipment, hazards of, 28
Fatalities:
  catastrophes, 40
  construction safety, 43
Fault Tree Analysis (FTA), 24, 151
Fed-OSHA, 117–119
Federal Drug Agency Center for Devices and Radiological Health (CDRH), laser safety and, 97
Federal Highway Administration (FWHA), 169
Fire safety:
  construction work, 46
  controls, 76–79
  cryogenics and, 51
  hazards, 75–76
  preparedness, 79

statistics, 74–75
Fireworks, 112
First aid, emergency planning and, 57–62
FOSHA, 120

Gallium arsenide laser, 96
Gas lasers, 96
Gaseous waste disposal, methods of, 173
Goretex, 91–92
Ground Fault Current Interrupters (GFCIs), 52
GULHEMP, back injury job-rating program, 33

Harnesses, fall protection, 72–73
Hazard analysis, methods of, 15–16
Hazardous materials:
  reference sources, 80
  types of, 79
Hazardous operations (HazOps):
  control, 86
  guide words, 84–85
  recognition of, 81–82
  review procedure, 83
  types of, 81–82
Hearing loss:
  agricultural safety, 28
  industrial hygiene and, 94
Heart attack, first aid procedures, 61
Heat Stress Index (HSI), 90
High Potential (HIPO), 16, 24
Highway Loss Data Institute (HLDI), 169
Human Reliability Analysis (HRA), 63
HyperCard, 198

Impaired workers, 68
Indoor air quality (IAQ), 89
Inductive reasoning, 154
Industrial hygiene (IH):
  air sampling, 87
  biological hazards, 88
  cumulative trauma disorders, 88
  environmental controls, 88
  epidemiology, 88

Industrial hygiene (*Continued*)
  ergonomics and human factors, *see* Ergonomics
  function of, 86–87
  industrial air quality, 89
  noise, 92–94
  thermal stress, 89–92
  vibration, 95–97
Industrial Safety Equipment Association (ISEA), 129
Industrial waste, types of, 170
Insurance Institute for Highway Safety (IIHS), 167

Job Safety Analysis (JSA), 16

Keep It Simple Stupid (KISS) training principle, 143

Laser safety:
  construction work, 47
  controls, 98–100
  hazards, 97–98
  lasers, types of, 96–97
Lawn mowing, guidelines for, 113
Liability, construction safety and, 48
Life safety:
  facilitating safe egress, 102–103
  safety egress hazards, 101–102
Lifting techniques, 34
Liquid-Cooled Garment (LCG), 91
Liquid lasers, 96
Liquid waste disposal, methods of, 171
Lockout/tagout systems, 30, 103–105
Logic tree perception, 20
Loss, defined, 11

Management Risk & Oversight Tree (MORT), 24, 140, 143
Management:
  function of, 176–177
  policy statements, 183
  safety professional, functions of, 178–179
  safety programs elements, 180–184
Man-Systems Integration Standards (MSIS), NASA, 65
Materials handling:
  agriculture industry, 30

back injuries, 37
construction safety, 46–47
ergonomics and, 68
safety system, 106–107
stress and, 32
welding safety, 174
Methanol-rhodium dye laser, 95
MIL-HBK-505, 152
MIL-STD-882 System Safety, 48
Mine and Safety Health Administration (MSHA), 107
Mining safety, 107–108
Mission Occupational Specialties (MOSs), 33
Motor Carrier Safety Assistance Program (MCSAP), 168
MSFC JA-4118, 152
MSFC-522A, 152
Multilinear events sequence (MES) perception, 20

NASA:
  fraction control systems, 151
  Man-Systems Integration Standards (MSIS), 65
  PRACA (Problem Reporting and Corrective Action), 147
National Electronics Injury Surveillance System (NEISS), 134
National Fire Protection Association (NFPA), 74, 101–102
National Institute for Occupational Safety and Health (NIOSH):
  accident investigation methods and, 23
  air sampling and, 87
  Work Practices Guide for Manual Lifting, 32–33, 37
National Safety Council (NSC), 27, 40, 177
National Traffic Safety Administration, 165
Neodymium yttrium aluminum garnet (Nd:YAG) crystal laser, 96, 98
Noise:
  industrial hygiene and, 92–94
  mining safety and, 108
Nominal hazard zone (NHZ), 98

Non life-threatening injuries, first aid
    procedures, 61–62

Occupational Safety and Health Act
    (OSHA), *see* OSHA
Office safety:
    emergency response, 117
    hazards and controls, 114–117
Off-the-job safety:
    controls, 110–115
    hazards, 109–110
OSHA:
    agriculture industry standards, 32
    catastrophes and, 38
    citation statistics, 122
    construction safety standards, 43
    defined, 117
    enforcement, 120–121
    farm machinery standards, 29
    FOSHA *vs.* State OSHA, 120
    function of, 117–118
    hazardous operations regulation, 83
    information sources, 122
    Permissible Exposure Levels (PELS),
        80
    policy statement guidelines, 184
    resources, 119–121
    standards, 118–119
    welding safety study, 173

Personal protective equipment (PPE):
    agricultural safety, 29–31
    body protection, 129–130
    chemicals and, 123–124
    cleaning of, 125
    cryogenic safety, 51
    ear protection, 127
    eye protection, 126–127
    face protection, 127
    foot protection, 128–129
    function of, 122–123
    hand protection, 128
    head protection, 125–126
    hearing protectors, 93
    laser safety, 98
    lawn mowing and, 113
    respiratory protection, 30, 95, 108,
        127–128
    reusability of, 125

selection of, 124–125
    skin protection, 30
    thermal stress, 92
    types of, 123
    welding safety, 174
Pesticides, agricultural safety, 28
Poisoning, first aid care, 114
Policy statements, 185
Pressure systems, 131–133
Probabilistic Risk Assessments (PRAs),
    63
Product safety, 133–135
Professional associations/organizations,
    4
Professionalism, 4
Public safety:
    checklist for, 135
    critical factors of, 136–137
    types of, 133–134

Radio frequency interference (RFI), 53
Raynaud's phenomenon, 94
Real Time Logic, 151
Recognition, significance of, 14–15
Recognition-evaluation-control strategy,
    13
Rescue breathing, first aid procedures,
    59–60
Resource Conservation and Recovery
    Act (RCRA), 55, 120
Risk, defined, 11
Robotics safety, 138–140
    control, 141
    hazards, 140
Rollover protection systems (ROPS), 28
RPTARS (Reliability/Problem Trend
    Analysis and Reporting System),
    149

Safety, defined, 9–10
Safety apocrypha, as reference, 199
Safety controls, hierarchy of, 17. *See
    also specific types of safety*
Safety costs, 4–5
Safety Data Requirements (SDRs), 185
Safety databanks, as reference, 197
Safety dichotomy:
    action *vs.* problem recognition, 13
    control, 16–18

Safety dichotomy (*Continued*)
  energy exchange, 15
  evaluation, 15–16
  recognition, 14–15
  recognition-evaluation-control
    strategy, 13
Safety documentation:
  analysis of, 188, 190
  change management (CM), 190–191
  significance of, 187
  specific, 187–189
Safety egress:
  facilitating, 102–103
  hazards to, 101–102
Safety engineering:
  basic concepts, 2–7
  defined, 2
  historical perspective, 1–2
Safety journals, as reference, 198–199
Safety manual, function of, 183
Safety philosophy, 5–6
Safety professional, functions of,
    180–181
Safety programs, elements of, 3, 182–186
Safety references:
  computerization, 193–198
  electronic databases, 198
  list of, 194–197
  safety apocrypha, 198
  safety databanks, 198
  safety journals, 198–199
  safety texts, 198
  trade magazines/digests, 199
Safety texts, as reference, 198
Safety training:
  defined, 141
  sessions, 142
  steps in, 142–146
Safety trends, 145–148
Schulmeyer Software Safety Risk, 149
Seatbelts, utilization of, 167
Sick Building Syndrome (SBS), 89
Single-event perception, 20
Sludge disposal, methods of, 173
Software safety, 149–152
Solid lasers, 96
Solid waste disposal, methods of, 172
Spinal injury, first aid procedures, 60
Sprinkler systems, 76–77

Staffing, safety management, 179
Stroke, first aid procedures, 61
Structural integrity loss, 153–154
Superfund Amendments and
    Reauthorization Act (SARA), 55,
    118
System, defined, 11
System safety:
  basic approaches, 155
  boundaries, 156
  defined, 11
  historical perspective, 154
  iterations, 163
  logic tree analysis, 158, 161
  matrix analyses:
    expanded, 163
    generally, 155–156
    plans, 156
    risk ratings, 161
    safety *vs.* reliability, 155

Tagout, *see* Lockout/tagout
Tandem Truck Safety Act, 168
Technique for Human Error Rate
    Prediction (THERP), 63
Terminology:
  accident, 10
  loss, 11
  risk, 11
  safety, 9–10
  system, 11
  system safety, 11
Thermal stress:
  cold stress:
    cause of, 91
    control of, 91–92
  heat stress:
    causes, 89–90
    control of, 90–91
    symptoms, 90
Time Petri nets, 151
Timing:
  accident investigation and, 22–23
  significance of, 6–7
Toxic Substances Control Act (TSCA),
    120
Transportation safety:
  car safety, 168
  control, 169

pedestrian, 169
human operator and, 167
motorcycles, 169
road systems, 168
seatbelts, 166–168
visibility, 167–169
Trench boxes, construction safety and,
    43–44
Trenching/excavation, 42–45

Unconsciousness, first aid procedures,
    61
U.S. Council on Aging, 33

Vapor cloud explosions (VCEs), 38

Vibration syndrome, 94
Vibration White Finger, 66
Video display terminals/units:
    ergonomic design and, 65–66
    office safety and, 116–117
Viton, 124

Walking and working surfaces (W/WS),
    170–171
Waste handling, 171–173
Waste treatment technology list, 174
Welding safety, 174–176
Wet Bulb Globe Temperature (WBGT),
    90
Work accidents, cost of, 4